Views from the Socially Sensitive Seventies

VIEWS FROM THE
SOCIALLY SENSITIVE SEVENTIES

Paul Lazarsfeld
Daniel Bell
Frank Graham
Philip Hauser
Suzanne Keller
William Ballard
John Kettle
Herman Kahn
Arthur Jensen
Fulton J. Sheen

Seminars Presented to the Supplemental Training Program of AT&T

International Standard Book Number: 0-88439-001-2
Library of Congress Catalog card number: 73-85003
Printed in U.S.A.

Copyright © by American Telephone and Telegraph Company, 1973

FOREWORD

The American Telephone and Telegraph Company's Supplemental Training Program has provided a forum for our people to discuss socioeconomic trends with the most outstanding experts in their fields. Educators, philosophers, futurists, men from business and government, all have attended to share their views with us. The program not only has made us more aware of the problems facing our society, but is helping us to understand how to contribute to their solution.

Reprinted here are ten S.T.P. sessions from the past several years which deal with what we term people-oriented problems. We selected these particular topics because we feel that people problems are among the most serious facing society today, and that solutions will demand nothing less than the best thinking of our greatest men and women.

This book is published in the hope that the opinions and information contained in it will reach a wide range of people and assist them in making the tough decisions which will be required of us all in the 1970's and beyond.

SAM E. BONSACK
Vice President
American Telephone
and Telegraph Company

TABLE OF CONTENTS

Foreword by Sam E. Bonsack — v

Introduction by Paul F. Lazarsfeld — ix

1. The Use of Social Science in Business Management — 1
   Paul F. Lazarsfeld
   Professor of Sociology
   Columbia University

2. The Post-Industrial Society — 10
   Daniel Bell
   Professor of Sociology
   Harvard University

3. The State of Our Environment — 30
   Frank Graham, Jr.
   Field Editor
   Audubon Magazine

4. Population and the Human Environment — 52
   Philip M. Hauser
   Professor of Sociology
   University of Chicago

5. The Future of the Family — 73
   Suzanne Keller
   Professor of Sociology
   Princeton University

6. The Survival of Mankind — 89
   William W. Ballard
   Professor of Embryology
   Dartmouth College

7. How to Think about the Future — 111
   John Kettle
   Writer and Consultant

8. Forces for Change in the Final Third
   of the Twentieth Century   132
   Herman Kahn
   Director
   Hudson Institute

9. Choices for Tomorrow   175
   Arthur E. Jensen
   Professor of English
   Dartmouth College

10. Revolutionary Philosophy of the Seventies   192
    Fulton J. Sheen
    Archbishop
    Roman Catholic Church

INTRODUCTION

My first exposure to the Supplemental Training Program at AT&T was as a guest lecturer. Since then I remained in touch with the program and became so impressed with its content that I encouraged the idea to publish ten of the presentations for broader consumption.

Eight of the talks reprinted here deal with what is called futurism. I am not a specialist in this area, but you will notice several sociologists among those who are. (By coincidence, the Russell Sage Foundation has recently brought out a volume on *The Sociology of the Future*.) Futurism is an amalgam of many disciplines, and this collection will be useful to university teachers and students in various subjects, including sociology, as well as to the business community. Certainly the social consequences of scientific and demographic developments are important to the staff of any corporation which plays a significant role in the contemporary social scene.

To catch brief glimpses of the way futuristic problems are viewed by a diversity of experts is very helpful. This volume includes the views of the professional futurists Daniel Bell and Herman Kahn, scientists such as the biologist William Ballard and the environmentalist Frank Graham, the demographer Philip Hauser, and the community sociologist Suzanne Keller. Arthur Jensen is a humanist who has been concerned with the activities of the "other culture"—the natural scientists. John Kettle is a journalist who has watched and reported on the activities of the futurists. Archbishop Sheen, who shares with the other contributors their concern about the future, emphasizes one possible remedy for its problems.

The utility of this volume will be increased if the reader is aware of the interconnections between the different papers. A number of basic ideas reappear in one or another form in each selection demonstrating that there is common ground in these experts' collective thinking.

FOUR BASIC FACTORS

Four factors appear as decisive elements in the futurists' discussions. One pair is basic: The numerical increase in the population and the continuous expansion of technical inventions. The other pair derives from the first: A growing population is bound to become more and more urbanized, and increasing technology leads to depletion of natural resources, or pollution of the environment. Obviously the four factors influence each other. All the presentations focus on the consequences of these four developments and their interaction, and on some of the proposed means of coping with them.

A mood of pessimism pervades the contributions: Most of the con-

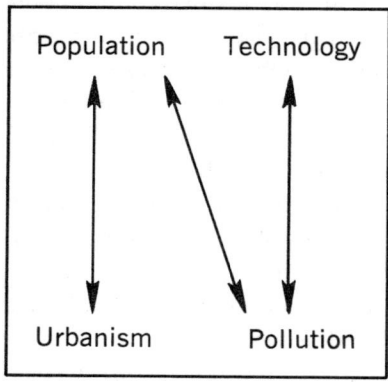

sequences seem undesirable, and few remedies seem to have been found. Moreover, the growth of the four factors has in recent years been explosive. We are dealing with the phenomenon called exponential growth. (Exponential growth can be simply illustrated by the process of doubling. If a set of 100 units doubles every 10 years, then in the beginning the growth is relatively slight: An increase of 100 units within the first decade, 200 units within the second. But during the tenth decade the increase will already be well above 25,000 units. Ballard, Kahn, and Kettle use charts to illustrate exponential growth in their subject areas.)

Population Explosion and Urbanism

The population explosion is dramatically described by Hauser. It took the whole history of mankind to attain a world population of 1 billion around 1850. The second billion was reached in 1930, and the third in 1960. By 1974 we can count on 4 billion. For the United States the explosion meant 200 million people in 1970 and probably 300 million by the end of this century Hauser also points out that the increase in urbanism can be documented in a similar quantitative way. In the United States alone, 65 percent of our people today live in metropolitan areas, and all further net population growth will be absorbed there.

Daniel Bell illuminates these statistics with a number of interesting details. For instance, adding one person to a city creates a very large increase in the costs of administering city services, while adding a person to a small town has no such drastic consequence. He also stresses the fact that the movement of the black population to the cities accounts for the rapid increase in visible unemployment figures. As long as blacks live on farms, however miserable, they are not counted as unemployed and are not included in the relief roles.

Technological Explosion and Pollution

Few of the contributors speak of conventional progress in science. Those who discuss the scientific explosion concentrate on its biological aspects. Jensen provides a useful list of biological inventions whose side effects were inadequately considered — such as certain antibiotic pills and insect sprays. He discusses other developments on the horizon which to him seem socially dangerous — such as drugs to duplicate people or to determine the sex of unborn children. In Jensen's view our difficulties are rooted in our belief that real progress can be made only by increasing and exploiting knowledge. He suggests that we should not go on to unveil nature's riddles just "because they are there."

Biologist Ballard points out that until now, man has been able to maintain his place in the ecological system by eliminating competing species and by artificially (technologically) increasing the available sources of food and energy. For example, when large-scale famine would have resulted if man had been forced to depend on the natural supply of food, machinery and fertilizers made it possible to increase the number of people who could survive. But the price was paid in natural resources. Ballard's presentation raises the questions: Should we seek slower technological and economic growth?

Graham, one of the editors of the *Audubon Magazine,* echoes Ballard's concern, although he understandably stresses the destruction of animal life. He points out three institutional villains. First are the legislatures. Second is the governmental bureaucracy. Industry is the third villain. His pages are an anguished review of environmental damage.

Finding Remedies

Why does it seem so difficult to stop these explosive trends? The various contributors mention at least three obstacles. One is the complexity of the social fabric itself. For example, Hauser discusses the connection between the level of education in a country and its willingness to accept birth control. If the population in a developing nation is growing rapidly, then few resources remain for education after the people are fed. Without resources not even a minimal educational level can be achieved, and so the people remain bound by tradition. In this situation they either cannot understand the technology of contraception or resist it for ideological reasons. So the population continues to grow, consuming the nation's resources.

Another difficulty comes from what are called "unintended consequences." Graham describes a chain of events which resulted from the efforts of the Audubon Society to save seagulls from extinction. At that time, hunters were killing them to sell their feathers to the millinery industry. A bill was finally passed prohibiting the shooting of gulls. But in the meantime the number of open garbage dumps had increased, and seagulls feeding on them, while being protected from their enemy the hunter, experienced their own population explosion. They began to prey on smaller birds, and as their numbers expanded, seagulls even endangered airplanes flying in and out of airfields near the garbage dumps.

Finally, the very speed of a trend can make it difficult to deal with that trend. Hauser points out that many political leaders grew up in small towns prior to the great urban explosion. These politicians retain a small-town value system, which limits their ability to understand and deal with the new, rapidly emerging urban problems. Their backwardness is accentuated by the fact that until recently, the election system for state legislatures gave the advantage to non-urban areas.

SOCIO-POLITICAL IMPLICATIONS

The anti-urban orientation of the state legislatures illustrates a major theme which runs through the selections in this book: the social and political implications of the four basic contemporary trends. Some of the talks given at AT&T went well beyond the stories one can cull from the daily newspapers.

The basic question these experts discuss is whether the problems can be dealt with by individuals or whether they require collective, organized decisions and efforts. A single individual cannot buy clean air or good transportation in the open market; such goals must be achieved by central agencies. But however the central agencies are controlled, their measures will by implication determine the price of

Views from the Socially Sensitive Seventies

1. THE USE OF SOCIAL SCIENCE IN BUSINESS MANAGEMENT

PAUL F. LAZARSFELD
Professor of Sociology
Columbia University

Can social science be of help to any kind of practical enterprise, to the farm, the city or the business organization? This is essentially what I will try to review with you. In the main, sociology is concerned with three problems: how the larger social units affect the individual; how we, as people, build up and maintain these larger units; and how people interact with one another. My examples will concentrate on the first problem, and I shall try to describe what is sometimes called the sociological perspective. You all have heard, I am sure, of the so-called depth psychologists. They try to find out what motives, often hidden, drive people to their actions — psychoanalysis is a good example. In a sociological study one wants to find out what social forces control a person — propaganda, coercion, subtle expectations or social values often unrecognized by the individual. Psychology sees men as pushed around from within, while sociology sees them as pushed around from without. Both, of course, see only part of the truth.

I shall not try to give you an outline of an introductory course in sociology, but rather to give you a few instances of sociological thinking. My hope is that each of you will be able to connect these examples with some of your own problems and experiences.

There are a variety of reasons why the impact of society at large upon us is not immediately obvious. Take, for instance, the field of demography, which includes the study of the age distribution of the population and how it changes. Demographers have traced various ramifications of the fact that people now live longer than they did a few decades ago; the implications are indeed strange. For example, fifty or sixty years ago, the death of a company executive was a major social catastrophe to his corporation, because he probably died while he was still working. Today, this same executive would have died during his retirement; his company would long before have planned for the reallocation of his particular responsibilities. Or, in sociological terms, his role will be filled by another without any drastic gap. (This is not to deny that there are always rare, creative executives who cannot in any exact sense *ever* be replaced.) But the point is that the change in life-span has forced our attention to the need for planned succession; it has also created new kinds of interpersonal tensions. For there is another kind of death in organizations. A vigorous man may lose his power almost overnight. It is predictable that at a certain point in his organizational career the executive will be regarded as socially dead. He is due to step down. This is an aspect of old age and of impending retirement that is not usually thought about. But what happens within the organization as the younger men begin to move in for the power struggle? This may sound unduly harsh, but sociologists of organizations have identified a constant series of conflicts within all types of companies.

The potential tensions are increased by another demographic change — the change in education. By now the majority of Americans can be expected to complete secondary school; in the foreseeable future at least half the population of the appropriate age will have graduated from college. This was certainly not the case twenty or thirty years ago. The result is that right now, young people are better educated, on the whole, than their seniors. Yet the older people occupy the power positions through their very seniority. As subordinates arrive with better skills than their superiors, a kind of generalized anxiety is created. Each faction is worried about or impatient with the other. The company will need to experiment with various types of solutions.

Of course these changes and the resultant conflicts are almost invisible to most of us. We simply take our situations for granted — until the social scientist stirs us up by exposing the social phenomena. Sometimes this makes us very uncomfortable.

I was struck by the fact that AT&T groups contain proportionally few blacks, Jews or women. Perhaps the organization has learned to take this for granted, but for an outsider, it is immediately a sociological fact whose implications are not altogether clear.

This type of analysis has led the sociologist to create some concepts which he uses as orientation when he tries to show the influence of society at large on a particular interpersonal situation. One such concept, for instance, is that of the reference group. Research has shown that most of the things we feel and experience are not really direct reactions but are relative to the groups we link ourselves to. For example, during World War II there was the famous study which asked soldiers their opinions of the draft. Offhand, you might have expected college men to have resented the draft more than blue-collar workers. College-educated men were forced to interrupt careers in which they had invested themselves; they had to take up new kinds of jobs. But the truth was that the college men were more likely to feel that the draft was fair, blue-collar workers that it was unfair. Why? Many workers in essential war industries were deferred from the draft; those workers who were called into the army regarded this disparity of treatment as grossly unjust. On the other hand, most college men were automatically inducted and came to feel—by comparison with their "reference groups"—that the draft reaches everybody.

This notion of the reference group appears in a large number of studies. I myself did one during the McCarthy period in the fifties to find out how college professors responded to the Senator's attacks on their colleagues. We were surprised to learn that the professors were fairly courageous. Usually professors as individuals are not likely to be heroes—that is not an essential qualification for the profession. Why were they so ready to take a stand? The answer became quite clear. They were more afraid of their colleagues than they were of McCarthy. The probability of being singled out by McCarthy as a Communist was relatively small, though it could, of course, cost a university job. But criticisms from a group of other professors were immediate, and few wanted to risk the danger of refusing to sign a petition or take part in a protest against McCarthy. Thus, punishment from one's immediate reference group was much more painful to bear than the real but distant danger of attack by a committee of inquiry. By and large, I would think that courage among soldiers is equally motivated by the desire not to let the group down. It is this group which controls subjective morale, even if external danger is objectively greater.

Let me give another example of an invisible social influence. There have been numerous studies of class differentials. One began with a simple experiment in which children were asked to draw a 25¢ coin. It was found that poor children will draw a much larger quarter than well-to-do children. A quarter was (at least at the time of this experiment) much more important to a poor child than a middle-class one.

Value was translated into size. Such studies are important because class differentials in attitudes and values have distinct consequences for many social organizations.

In our country, we spend a good deal of public money on health clinics, hospitals, and nurseries, but they are usually underutilized by the intended clients. Thus people in the ghetto districts do not take advantage of health facilities to the degree that was expected. Why? According to sociological research, there is a good reason. There is a distinct contradiction between the bureaucratic organization of a well-run hospital and the anti-bureaucratic orientation of underprivileged populations. In a good hospital, activities are well organized under separate jurisdictions. There are entry procedures, there are forms, there are schedules. A general examination precedes any treatment by a specialist, and particular complaints entail a series of visits. The ghetto dweller is invariably frightened by this formalism; he retreats to his family and simply doesn't come back. The difference between a police station and a hospital is not so obvious to the newly arrived Puerto Rican as one might think. Both are formidable institutions where the individual feels lost. In recognition of this attitude, some communities have developed "storefront hospitals" where there may be only one nurse or doctor, where procedures are simple, where the usual hierarchy is minimized. Perhaps such a hospital is not up to the standards of modern medicine, but it is far more acceptable to the patient who, as we now say, comes from a poverty culture.

Our society pushes all of us in hidden ways, not just those who are members of some particular socioeconomic class. Sociologists who have studied American business have become very aware of the socially induced tensions which permeate our lives. For instance, have you ever put together all the proverbs that urge you to love your neighbor but to work hard and make a profit? This is not easily accomplished. The proverbs say that the early bird catches the worm, but they also say that you shouldn't hustle and rustle all the time. Just as these social proverbs are contradictory, so are many of our institutional arrangements. One Bell Telephone study on supervision of operators found that the supervisor has two incompatible functions. She must teach and advise the operators, but she must also rate their performance. Consequently, the girls are hesitant to ask questions about their jobs because they are afraid to reveal weaknesses and lower their ratings. The company has created a tension between the functions of rating and of helping. The same is true of the job of social work. The caseworker is supposed to help the man on relief, but at the same time, she is policing the use of public funds. There is a definite inconsistency between the policing and therapeutic functions. You find the same contradictions embodied in many organizational roles, and they are not restricted to American businesses. In France, for example, the importance of the family is in some ways much greater than in our country. Americans generally accept the idea that men should be appointed and promoted according to their individual merits. But in France the family is so dominant that if you have a nephew or a cousin who wants a chance in the family business, you appoint him divisional supervisor of one of your factories. You do this solely because he is your nephew, irrespective of any of his own abilities. Thus the Frenchman is faced with a continuing contradiction between his responsibilities to the family and his obligations to the business. As a matter of fact, some sociologists have recently studied the problems of the family business and how it survives.

Social scientists certainly don't have all the solutions for these problems. Thirty or forty years ago, for example, they recognized that employees want to be able to express themselves through their work, to realize their human potentialities. This led to quite a long tradition of human relations studies and of many changes in the mode of supervision, particularly supervision in factories. No longer was the efficiency of the worker the employer's only concern; he had to take into account the worker's feelings, his desire to participate in company decisions, his attitudes toward fellow workers, etc. Now all this has become controversial. Some social scientists have dared to ask how we know that people want to fulfill themselves in the factory. Perhaps this is the social scientist's attitude, not the worker's. Maybe the worker just wants to get the job done as quickly as possible and turn his attention to his own leisure-time pursuits. Perhaps many jobs can't be made more humane anyway, particularly with the advent of the computer. So instead of humanizing the work, let's shorten it. But others ask, What will the consequences be? Leisure time wasted on meaningless activities? The idea that work destroys the soul has evolved into the suspicion that leisure does. There is some evidence that this is not an empty idea.

A case in point is a technological matter very much in the public eye. If free time is used for television watching, can this become an addiction, like alcohol or drugs, where the individual no longer has the power to control his behavior? Social scientists have tried to find out.

One study used photographs of people reading, people sitting at a bar, people watching television. Seven or eight of these photographs were shown to about 3000 people, who were then individually interviewed about their reactions to the activity in the picture. From a list of fifteen statements such as "This is very interesting," "I will regret it tomorrow morning," etc., they had to choose one that expressed their evaluation of the pictured activity. As you might expect, reading took first honors, and most agreed that drinking was shameful. (You may know that the United States has a relatively high illiteracy rate, but we are conditioned to think that reading is self-improving.) Television watching, on the other hand, elicited ambivalent reactions. The people interviewed mentioned its involuntary aspect, the viewer's tendency to keep watching in the hope that the next program would be better.

Social scientists have invented a whole new activity for the individual businessman—the well-known sensitivity training. I'm sure you have all heard how it has swept the country. Business organizations send their managers to special T-group sessions that will help them gain skill in interpersonal relations. There is the psychodrama technique, where the participants must play various unaccustomed roles. The boss takes the part of a stenographer, for example. There is the discussion technique, where problems that arise in the course of a day are talked through with the help of a trained leader. It is very difficult for me to say whether these are legitimate social science activities. Certainly they have many therapeutic effects. One thing that particularly strikes me is the gradual crumbling of what we in our jargon have called "pluralistic ignorance."

Many people believe that they have troubles no one else shares. Fifty years ago, a former collaborator of Freud's named Alfred Adler came out with the heretic statement that the social world is not controlled by sexual feelings but by inferiority feelings. This was a surprising statement at the time, because individuals had assumed that they were alone in feeling inadequate. Today we assume that everyone has some kind of inferiority feelings.

Pluralistic ignorance—or unfounded assumptions—affects many social situations. For instance, if you were to study whether relatively intelligent Northern whites would seriously object to having black neighbors, you would probably find out that they would not. But the same intelligent Northern whites are positive that everyone else would object to it. This is a variation on pluralistic ignorance—we don't think that the other person feels as we do—and we are trapped into inaction.

Since sensitivity training maneuvers the individual into a situation where he has to listen to others, he may suddenly discover that he is much more like others than he had dreamed. This finding has all sorts of implications for human behavior.

In one study, students were asked to come to an experimental session and were then told that the experiment involved some electrical equipment that would hurt them a little bit, but not much. As a result of this talk, the students were badly frightened and uneasy about the equipment. The investigator explained that the equipment wasn't working well, that the students would have to wait for half an hour while it was being repaired. They were given the choice of waiting with the group or of going to the library to read. By and large, students decided to wait together. From subsequent interviews about this anxious waiting period, it became clear that they wanted to be with others in order to find out whether the others were just as scared as they were. They couldn't ask directly, "Are you scared?" because they were ashamed to, but they tried all sorts of indirect tactics to "sense each other out."

This "sensing out" is not purely an experimental by-product; it is very important in many organizational situations where the individual cannot really ask, "What are the unwritten rules here?" "Can I tell off-color jokes?" "How close to 9 o'clock must I arrive in the morning?" These underlying details of social life—norms, as the sociologist calls them—must be discovered, sensed out, in order to counteract feelings of ignorance or discomfort. Not knowing what other people do or feel has led to a variety of mechanisms by which people find out about each other.

I hope that we are better able to find out about one another, and that the chairman will give you a chance to ask questions. Thank you very much.

Paul F. Lazarsfeld

Paul F. Lazarsfeld is a sociologist with the unique ability to relate the social sciences to the complex areas of industrial and business management. Since the advent of the socially sensitive seventies, this talent has become even more valuable to large organizations and businesses than it has been in the past.

A member of the Columbia University faculty since 1939, Professor Lazarsfeld is today the Quetelet professor of social science at Columbia and chairman of the board of the Columbia Bureau of Applied Social Research. Born and educated in Austria, he holds a Ph.D. in applied mathematics from the University of Vienna. A few years after he received his doctorate, his interest moved toward applied psychology and he became director of the Division of Applied Psychology at the University of Vienna. In 1933 a Rockefeller Foundation grant to observe techniques of psychological research brought him to the United States, where he has lived ever since.

In 1937 he became director of the Office of Radio Research, established by the Rockefeller Foundation to study the effects of radio on American society. When the office was incorporated into the Columbia Bureau of Applied Social Research, he became the bureau's director. For many years Professor Lazarsfeld continued to research and write about radio, and then about all media of mass communication. He established a national reputation as a public opinion researcher and was especially well known for his application of statistical means to the study of relationships between mass communications and American voting habits.

During World War II, he was consultant to the Office of War Information, the War Production Board, and the War Department. He advised these professional groups on the role of American public opinion in the shaping of foreign policy.

In 1949 he resigned as director of the bureau to devote more time to research and writing of a highly specialized nature. He pointed out that social science research needs a more precise language of measurement and that the use of math in the social sciences should be expanded. Since the early 1960's he has focused on mathematical concepts of sociology.

Professor Lazarsfeld has served as president of the American Sociological Association and has received many awards and honors, among them an honorary degree from the University of Chicago. He has written a number of books and has been widely published in professional journals and general magazines.

BIBLIOGRAPHY

and Samuel Stouffer, A. J. Jaffee. *Research Memorandum on the Family in the Depression.* New York: Social Science Research Council, 1937.

and Hadley Cantril, Frank Stanton. *Radio and the Printed Page.* New York: Duell, Sloan & Pearce, 1940.

and Frank Stanton. *Radio Research, 1941.* New York: Duell, Sloan & Pearce, 1941.

"The People's Choice": How the Voter Makes Up His Mind in a Presidential Campaign. New York: Duell, Sloan & Pearce, 1944.

and Frank Stanton. *Radio Research, 1942-1943.* Essential Books, 1944.

The People Look at Radio. Chapel Hill, N.C.: University of North Carolina Press, 1946.

and Patricia Kendall. *Radio Listening in America.* New York: Prentice-Hall, Inc., 1948.

and Frank Stanton. *Communications Research, 1948-1949.* New York: Harper & Row, Publishers, Inc., 1949.

and Robert K. Merton, eds. *"Continuities in Social Research": Studies in the Scope and Method of the American Soldier.* New York: Free Press, 1950.

Mathematical Thinking in the Social Sciences. New York: Free Press, 1954.

Voting. Chicago: University of Chicago Press, 1954.

and Morris Rosenberg, eds. *"Language of Social Research": A Reader in Methodology of Social Research.* New York: Free Press, 1955.

and Wagner Thielens. *"Academic Mind": Social Scientists in a Time of Crisis.* New York: Free Press, 1958.

and Elihu Katz. *"Personal Influence": The Part Played by People in the Flow of Mass Communications.* New York: Free Press, 1964.

Uses of Sociology. New York: Basic Books, Inc., 1967.

and Neil W. Henry. *Latent Structure Analysis.* Boston: Houghton Mifflin Co., 1968.

and Neil W. Henry, eds. *Readings in Mathematical Social Science.* Cambridge, Mass.: M.I.T. Press, 1968.

2. THE POST-INDUSTRIAL SOCIETY

DANIEL BELL
Professor of Sociology
Harvard University

When I came in this afternoon I was greeted with a number of cartoons which heralded my talk. One of them which you may have seen shows someone looking out toward a rising sun on the horizon labeled Post-Industrial Society. It reminded me of an old story of a Communist agitator in the Soviet Union who would go out and talk about the future and say that communism is just on the horizon. Then somebody told him that a horizon is an imaginary line which recedes as you approach it. I'm not so sure that the post-industrial society will be an imaginary line as one approaches it.

Actually what I'd like to do this afternoon is talk about more than the post-industrial society. I'm interested in social forecasting. I don't believe that the future is a point somewhere abstractly out there in time. The future unfolds from structural tendencies in the present. And what I'd like to do is talk about American society thirty years past and thirty years hence, because what comes up in the future is not a complete surprise. It's not inevitable either; but it derives to a considerable extent from changes in the society.

Now, if one were to do a detailed picture or an analysis of social change, one would have to deal with about four components: changes in values, changes in culture, changes in politics and changes in social structure.

The problem of changes in values is quite difficult, because values unfold slowly, or sometimes have sudden breaks. Culture is more volatile today than ever before—and what's more important, one finds different life styles among different groups. It's much harder to chart that.

Politics is probably the most difficult to chart, and I shan't go into that area. What I'll be talking about primarily is changes in social structure, with due appreciation, I trust, to the fact that I'm not trying to sketch a picture of total change in the society, but change in one dimension of four: values, culture, political structure, social structure. And by social structure I mean primarily changes in technology, demography, the economy and the socializations which derive from them.

METHODOLOGY OF FORECASTING

If one is trying to do forecasting, as I have been in the past several years, there's a very important methodological difference in the kind of forecasting one does, let's say, in a sociological framework compared with other areas. It seems to me there's an important principle involved in the way in which people try to look ahead at a society. I can explain it best, perhaps, by a number of comparisons.

Technological forecasting has a certain simplicity (although there are always many surprises). One is dealing primarily with physical parameters of fairly recalcitrant objects, whether it be speed trend curves, strength of materials, or so on. And one can have, by now, fairly well-defined procedures. It might be a morphological technique of Fritz Zwickey, which is simply the infinite permutations and combinations of the parameters of the variables of a problem you're dealing with. Or it might be an envelope curve technique which Robert Ayres and various others have pioneered, which again takes into account the finite physical parameters of the particular problem one is dealing with. So to that extent it is possible to do a certain degree of technological forecasting in a painstaking way.

Economic forecasting is quite different. There are two kinds: the economist deals primarily either with trend series or with particular problems. With a time line he can project agricultural productivity or manufacturing productivity, or whatever other time series he wants; or perhaps in a more sophisticated way he can create a model of the economy using an econometric model (there are several today). But you have, again, fairly well-defined variables and you can chart, again, their permutations and combinations, and work out a model on this basis, such as a Brookings model, or the University of Michigan model —various others of this kind. You have a model of the economy and deal with it, and again have a certain degree, at least, of elegance and precision.

Political forecasting I would regard as the most recalcitrant of all, and here I can simply dispose of it with one anecdote, which is what

I call the formulation of Brzezinski's law. This is named after a former colleague of mine at Columbia, who is the research director of the Institute of Communist Affairs at Columbia. He was once being baited on television by a commentator who said: "You're a Kremlinologist." The professor said "Yes." He said, "You chart the power struggles of the Soviet Union?" He said, "I try to." The commentator said, "You advise the State Department of your findings?" He said, "When they listen." "Professor Brzezinski, how come you failed to predict the ouster of Khrushchev?" Brzezinski's very quick in speech; he said, "Tell me, if Khrushchev couldn't predict his own ouster, how do you expect me to do it?"

Now, this is one of the elements, it seems to me, of political forecasting. It is primarily a problem of intelligence, in the very literal sense of the word; namely, inside information about the actual play of power which goes on in crucial situations of this kind.

Social forecasting is different in many ways from the other three. It's not, in a sense, as volatile, as erratic as political forecasting. It's certainly nowhere near the kind of model you can get in econometrics, or the fixed physical parameters, if you will, of technology.

And what at least I think is the way to do social forecasting — perhaps forecasting is a misnomer here — is essentially this: to try to identify relevant social frameworks — not, as many of my colleagues do, to look at a particular item, such as leisure or hours of work, and simply speculate about particular items or particular traits or particular trends. I distrust most of the projections which deal with particular items (such as: In the future man will work a twenty-hour week; or People will live here rather than there; et cetera), because you have no way of knowing what's going to hang together with what. And therefore the projection of particular traits is very deceptive. It's also very seductive; it sounds very nice, very fine. A more difficult and perhaps less satisfactory way is to say: Can you identify relevant social frameworks and try to see what hangs together with what, and then deduce certain problems? Therefore social forecasting is not primarily the projection of trend lines or series. It's more the deductive process, and it's the deductive process based upon some fairly sophisticated way of trying to see what hangs together with what in a society. To this extent it depends more than any others on the process of conceptualization and reconceptualization of items.

Now, there's a lot of jargon in much of the social science methodology, and I suppose "conceptualization" is part of it. By conceptualization I mean something very simple. I mean a process of being fairly self-conscious about why you select the attributes of a complex phenomenon, and why you group them together the way you do for purposes of analysis.

To be very simple-minded, if you were to say to a four- or five-year-old child, What is it that a whale and an elephant have in common? he'd say, They're both big. In a sense, even though one swims and the other walks, he's taken an attribute, size, and grouped them together for purposes of comparison. Size therefore is a conceptual attribute.

If you were to say to a bright twelve-year-old, What is it that a whale and an elephant have in common? he'll say, They're both mammals — even though one swims and something spouts out from his top and the other has tusks and walks in a lumbering way. They have a common reproductive structure. They have a common respiratory structure. Again he has selected out certain attributes, things which are highly disparate in the phenomenological sense, and grouped them together for purposes of comparison.

That's all conceptualization means, really; a self-conscious knowing why you've selected attributes of a complex phenomenon. All of knowledge is a process of selection, and the sophistication lies in knowing why you've selected particular attributes and grouped them together in the way you have in order to deal with them in this respect.

One of the case studies of the Harvard Business School illustrates this. About thirty years ago, somebody went to the Cunard people and said: "Why don't you go into the aviation business?" They said: "Why should we go into the aviation business? We're a shipping line." They said "you're not in shipping; you're in transportation. You have people who have offices in different cities. You know the problems of customs, of routes, of travel, et cetera. You're not in the shipping business. You're in transportation. And if you're in transportation, you ought to follow the signal and go into aviation." They said: "We're a shipping line. All we know is ships. We don't know anything about planes."

The result is that the Cunard shipping line today primarily consists of one ship in Long Beach, California, and one in Fort Lauderdale, Florida. And that's the end of the Cunard shipping line.

What I'm going to do this afternoon is look at some familiar facts about America and try to present some reconceptualizations. First, I'm going to do a quick sketch of the demographic picture of what's happening in society. Second, I'll offer three conceptualizations which

I will label in the schematic way: the creation of a national society, a communal society and a post-industrial society. What these will be is an effort to take certain familiar facts and, by grouping them together, perhaps in a novel way, deduce certain things which normally might not have been noticed.

DEMOGRAPHY

First, the demographic problem of this country: it seems to me there are many things one can illuminate by simply understanding a set of basic figures. If one takes a look at what's been happening since 1945, many of the problems we face come from sheer magnitudes and numbers. They don't come wholly from ideologies. They don't come from stale notions about creeping socialism or fading capitalism, or anything of that sort; they come from the impact of numbers.

The most important fact is that since 1945 there have been 90 million births in this country, and, subtracting for the deaths, a net increase of 60 million: from 140 million to 200 million. In a single decade from 1950 to 1960, 28 million people were added to the population. This ten-year increase is as large as that in the seven decades, from 1790 to 1860, from the formation of the Republic to the Civil War.

In percentage terms, the increase may not be so large as in the early years. In the early years of the Republic, we moved from 5 to 7 million, which is an increase of 40 percent. And the post-war increases have been at about the rate of 20 percent per decade.

Well, this impact of numbers comes within two different contexts, both of which are crucial to the nature of what's happening in the society. First there's the change of scale. If there's anything which is important (and I'll come back to this later) about the nature of society, it is this. Few things are wholly new in society. Few things are wholly novel in the history of human imagination. Most things which go on today have been thought of before and even done before. What is novel is a change of scale; the increase in numbers of people who're doing it, and the problem of management as a result of this.

The change of scale comes on top of something else, which is a change in institutional form. The early persons who were added to the society were added in a segmental fashion. They were simply extended as links in a chain across a continent. The additions which came in the last quarter of a century were pyramidal. They landed on top of existing concentrations. To the extent that they made a previous concentration even denser, they added new burdens. They also arrived in a high-consumption society in which people will not put up with what they might have accepted thirty years ago. My father, who came over as an immigrant in 1910, certainly did not have the kind of expectations anyone has, let's say, who comes in from the South today, or who comes from a rural background.

To accommodate these changes in scale and expectations, a complex infrastructure must be built. And if one considers the fact that today it costs about $18,000 capital per person to provide infrastructures in the society — that is, water, sewerage, schools, housing, hospitals, et cetera — and if you add 60 million people, you see the kind of financial burden which societies have had to face in the last quarter century.

There has been not only an increase in numbers, but changes in the way in which these numbers come up. From 1950 to 1960, for example, the number of young people (14 to 24) was almost constant. From 1960 to 1970 the number of young people jumped 44 percent. Looking at this fact helps to explain some of the agitation which has been going on in the schools.

I don't want to minimize the political events, the ideological events, the consciousness of a war which has been damaging to the morality of this society. But you also have the fact of a sudden 44 percent bulge in the number of young people coming into the society. And therefore there is an extraordinarily heightened competition for place and position.

Though it has usually been overlooked, there's another important aspect to what's been happening in this country in the last quarter century. We think of ourselves as very largely an urban society, and as having been so for a long time. But actually the steepest change in our movement from a rural to an urban society occurred from 1940 on, even though a majority of the people had already begun to live in urban areas by about 1910. But it's only since 1940 that we began to get a rapid and concentrated passage of people out of agricultural areas. This comes from the fact that from about 1910 to 1940, agricultural productivity in this society was growing very slowly — steadily, but slowly. In thirty years it went from an index of 100 to an index of 125. From 1940 to 1960, agricultural productivity jumped from 125 — using an index, again, of 100 in 1910 — to 400. This was an increase of 375 points in twenty years. It was largely because of the huge demand generated by World War II, but also because of the huge increase in new fertilizers, plus some new machinery. We paid for the productivity boost in a very different way. Much of the pollution in our rivers is not due to industry; a large portion is due to nitrates, which are run-offs from the fertilizers of farms.

Because of the sudden jump in agricultural productivity, 25 million people left the farms for cities from 1940 to 1965. In the 1940's the out-migration was 1.3 million persons a year. In the 1950's it was down to 1 million a year, and by the 1960's it was down to ¾ million persons per year. By 1965 there were half as many commercial farms as in 1940. In the same period the number of farm tenants dropped from about 600,000 to 82,000—an extraordinary shrinkage.

Now, one of the crucial elements in this change was that proportionately, the large number of persons who left the farm were black sharecroppers. The large number of Southern blacks who had always been rural have become highly urbanized. And this fact is responsible for an important development which is hidden in the way in which people understand statistics. In the 1960's we heard that the black unemployment rate was double that of the white, which is true. But the implication was that somehow blacks had been working before and now were suddenly unemployed. And that's not true. In any basic type of economic analysis, as most of you know, society is divided into a household economy and a market economy. A household economy, by and large, is not figured in any kind of GNP data or any kind of statistical analysis. If a man is on the farm, even though he's doing very little, he's not counted as unemployed; he's simply an extra mouth to feed. If he moves to the city, he *is* counted as unemployed, so the unemployment figures go up.

The real problem of population is no longer simply numbers, but, obviously, distribution. Population densities become in many respects the crucial question, particularly densities in some of the larger metropolitan areas. But densities are subject to a very important new consideration: the fact that the location of work, the location of industry, is no longer dependent upon the kind of market force that it had been before. Historically, most cities were located near waterways and raw materials. The basic heartland of this country—which is essentially the Chicago-Pittsburgh-Cleveland-Buffalo-Lake Superior area—developed because there was coal in the western Pennsylvania region, northern Kentucky, southern Illinois; iron ore in the Mesabi range; and a river system to hang it together.

But in a service society—which is what we begin to have today—the location of industry is no longer dependent, or at any rate is less dependent, on raw material forces or even market forces. It becomes subject to more conscious choice, at less cost, in this respect. The largest growth industries in the country are primarily health and education, as a major component of service. The location of new schools, the hospital complexes, are no longer dependent on their market forces. And if it were based on more planning than we have ever had before, a national population policy could redistribute population to a considerable extent.

THE NATIONAL SOCIETY

So much for demography. Let me quickly turn to three notions, or conceptualizations, if you will, as we try to understand what's going on in American society.

The first is the idea of a national society. Now, we've always been a nation. In some respects we were the first new nation, because ours was the first nation created out of a revolution with a constitution as a written act. We've always been a nation as a psychological fact. We became a national government during the famous Webster-Hayne debate in 1830, when South Carolina objected to paying taxes to support projects in Ohio. We decided we were part of a common union, a decision which was reinforced by the Civil War; and so we acknowledged national government.

In the New Deal we had to acknowledge the fact that we had become a national economy, and that basically the growth of national markets and the growth of national corporations meant the need for regulation. Therefore we began the self-conscious introduction of devices for the regulation of the national economy: the SEC to regulate securities markets, the NLRB to decide labor unit bargaining—devices of this sort.

We have become a national society in two fundamental respects. One is that the basic problems of health, education and welfare can no longer be handled on the basis of the units—the states—which the Constitution had assigned to handle them. The states have inadequate tax bases. One finds today that health is becoming more and more a federal concern (Medicare and Medicaid). The elements of education are now funded by the federal government through the Elementary School Education Acts, the Secondary School Education Acts and the Higher Education Acts. Welfare becomes the national responsibility, even though Congress has so far refused to accept the proposal put forth by President Nixon.

And we've become a national society not only in the sense that key elements of the quality of life—health, education and welfare—have now been moved to a federal center; but also in the sense that we've become more integrated, largely through the media and through the revolution in transportation and communications. This has meant the beginnings of a national awareness. So that anything which takes place in one part of the society has immediate repercussions in every other part.

Now, if you look at the fact that we've become a national society, a number of things follow. One is the fact that the structure of fifty states makes no sense; that the existence of New Jersey or Delaware or Rhode Island, with contours which are simply historically based, has no meaning in any respect. And the kinds of burdens assumed by state government have no meaning.

A few years ago when Governor Hughes was inaugurated as the governor of New Jersey, he made a speech which I thought had been written by a Harvard undergraduate—and I mean no disrespect. He said: "New Jersey is suffering from an identity crisis. Here we are caught between Pennsylvania and New York. Half our people go to work in Philadelphia and the other half go to work in New York." But what is New Jersey? What is the meaning of fifty states in terms of the problems of our society?

One thing which gives people a sense that their society functions is the efficient handling of services, whether they be mail services or telephone services or garbage collection. People have to have this sense of effectiveness, which depends in large measure upon rational administration.

Ours is a very modern economy with a high-grade technology. Yet it is a Tudor polity, and I mean it in almost a literal sense . . . a structure of counties, townships and sheriffs. What do you need a sheriff for? There's no Nottingham Forest around any more. Yet we have sheriffs.

The New York metropolitan region, for example, has fourteen hundred local governments, but that's not decentralization—that's disarray. There are water districts, health districts, sewerage districts, park districts, police districts; many with taxing powers of their own . . . and no rationale for who is to do what. In a basic sense, if we have a national society, one of the fundamental challenges is to figure out what is the appropriate social unit of what size and scope to handle what kind of problem. What is to be done on a local basis? What on a neighborhood basis? What on a township basis? What on a metropolitan basis? What on a regional basis? What on a national basis?

So one of the fundamental problems, I would argue, in the next thirty years, is the need to find new administrative mechanisms. This can be the most difficult of all, simply because you have the recalcitrance built in by a number of office holders and the political machines who benefit from the existing archaic structure.

In New Jersey there is something called the Mosquito Control District, which has enormous powers; so enormous that they're able to rake off half a million dollars a year or more in Mafia-controlled rackets, shaking down contractors who don't have to satisfy the requirements of the Mosquito Control District.

And it turns out that in some areas of the country you *elect* people to Mosquito Control Districts. It turns out that in Minnesota there are fifteen different levels of government which people vote for. One of the fundamental problems of disorientation that people feel in this society is the failure of a whole series of administrative mechanisms—the failure of the political structure to accommodate to the changes in the nature of the social structure itself.

So making a political structure congruent with the social structure is one of the crucial problems of the next thirty years, and it looks like an almost impossible one.

One of the other problems which arise from the fact that we have a national society is that all political decisions now get funneled into a specific cockpit, which is the national government in Washington.

American society has escaped some of the internal wars which have plagued the large part of Europe because of the insulation of space. There's probably been more violence in this country—by any rough set of indicators you choose—than any country in Europe, particularly labor violence. It's hard to get a statistical measure, but to the extent that one can, these indicators would be number of strikes, number of times troops have been called out, number of man-days lost, number of persons killed. But much of this has taken place at the perimeters of society: in mining camps, timber camps—and it was a long time before it had an impact on the political center.

In 1893 there was as severe a depression as ever happened in this country in terms of proportional impact upon persons, businesses that went bankrupt, and number of people unemployed. An army marched out from Massillon, Ohio, headed for Washington to demonstrate. It started out with 10,000 men, led by General Jacob Coxey. By the time they got to Washington there were a few hundred ragtag and bobtail people left, and the phrase Coxey's Army passed into the language indicating something which is kind of silly.

In 1963 Martin Luther King and A. Philip Randolph called for a march on Washington, and in 48 hours there were a quarter of a million people descending on the city and mobilizing pressure in this direct way. Since then you've had peace marches, Pentagon marches, et cetera. And it's quite likely that in the next thirty years you'll have more and more of this kind of mobilization politics as a basic form of pressure on the central government And this, too, it seems to me, is a very important change in the nature of the political structure of the society.

THE COMMUNITY SOCIETY

My second notion is that of the communal society, which refers to the nature of claims on a society. There are three dimensions of a communal society: a rise in non-market public decision making, a claim of group rights rather than individual rights, and an increased desire for participation in the society.

In the very simple sense of economics textbooks, goods are of two kinds: individual goods which you adjust to your own taste, or social goods which cannot be bought by a single individual. A suit is a consumer good of individual taste; you buy it to fit your own frame. A gun is not—although in 1789 when you went into the army you brought your own musket. Today you don't; your rifle is a public good purchased by government, and therefore standardized in terms of the equipment the government provides.

In the same way, some purchases which have to be made today, particularly environmental purchases, are essentially public goods. Nobody can buy his share of clean air in the marketplace. You can exempt yourself from some of the hazards by purchasing humidifiers and filters and air-conditioners. But if you go out into the street, you can't buy your share of clean air. More and more the framework of the society in terms of roads, housing, environment, is made up of public goods purchased through public mechanisms, whether it be by the federal government, state, local or what have you.

Now, this leads us to a very important consideration. The virtue of the market—and there are many, many virtues to the market—is that it tends to disperse responsibility. The market is truly, in the Adam Smith sense, an invisible hand. You don't know who to blame for what goes on in this society. If a firm fails to meet a shift in taste, it has only itself to blame, and goes broke.

In the 1950's the industry which lost proportionately the single largest number of jobs was the textile industry. Apart from the element of foreign competition and certain regional shifts, the most important reason was the decrease in the average age of marriage. You say, what does this have to do with the textile industry? And that's part of the beauty of sociology: to underline these relationships. Before, a girl married at the age of twenty-four; now she marries at nineteen. Before, a boy married at the age of twenty-six; now it's twenty-two. It means less dressing up, less going out, fewer purchases of suits, more purchases of durables, houses, and automobiles. The result of this decline of the marriage age was a sizable loss of jobs in the textile industry. But this is a change registered through the market, and there's nobody to blame.

However, a non-market public decision is visible to all. You know who makes the decision. It's made in city hall, and you know whose ox is going to be gored, and you go down and fight. The result is a road that goes through a ghetto or a road that goes through a rich section of town. Or it's a jetport in New Jersey, in the Meadows, or any other place. Everybody knows who's going to have to pay some of the cost. To that extent, the very nature of non-market public decision making increases community tensions and community conflicts.

And the sorry prediction is that in the next thirty years there'll be more community conflict in America, not because people are more cantankerous, but simply because there's a shift in the locus of decision making. The shift is away from the market to non-market public areas. Along with it goes that situation which is allied with the term used by economists, called externalities: the fact that more and more decisions reached by private parties have third-party effects. So that when a firm uses a river to dispose of its waste, it generates a cost, and the question is: How do you decide who's going to bear the cost? And, again, a public agency is going to have to make that kind of decision. So you increase the power of public agencies, and you increase the tendencies toward conflict.

A second aspect of the communal society is the fact that you have more and more attention to group rights. Let me take the problem of the blacks, because they are now levying claims on the society based on a historic disadvantage. What they want is a group right, quotas, preferential hiring and various kinds of advantages given to them as an attribute of their color, by virtue of the fact that they're a group. Now, if this were a problem of right versus wrong, there would be very little question of how to get justice in a society. But in any society the problem is not right versus wrong, but right versus right.

The case of the New York City schools is a perfect illustration. Sixty years ago, if you wanted to become a school principal, you had to know the leader of the local Tammany Club. That kind of political influence was what got you the post. In an effort to change the situation, a merit system was introduced. The merit system meant that you had to become a principal on the basis of certification by examination; you needed certain educational qualifications, as tested by the board of education itself. So a merit system was substituted for a patronage system. But now you have blacks coming along and saying: We want black principals, even though they may not be on any list. Because where you have a large number of black children in a school, a black principal is to be a model. There may be some claim of truth in that kind of argument. But then what do you do? Do you take a man who's worked for twenty years, gotten himself a higher education, come out first on the list, and simply say: We bypass you, even though you have the merit? There is no real answer according to any principle of abstract justice, because the claim here is one of right versus right, not of right versus wrong. And the claim of group rights increases tension in a society.

The third point about the communal society is the demand for participation. Increasingly people resent the fact that they have no voice in making decisions. But the demand for participation goes along with another kind of demand, which is for community control. The problem with community control is that it increasingly segregates a society. If you turn over more and more aspects of schools—let's say, neighborhood issues, zoning, et cetera—to complete community control, it allows segregation by ethnic group, class group, color group, et cetera, in a way which the society has never had to face before. Curiously enough, those who were segregationists in the past or racists down South and wanted to keep blacks out of their schools would have the same claim today on the ground of community control.

What happens in a society which has emphasized the notion of a common learning, of a common citizenship, when confronted with the problem of segregation based upon community control? How do you balance the rights of participation against those of common citizenship? Do you allow particular groups who live in a community to be veto groups over a society?

Increasingly in this society the politicalization, if you will, of decision making focuses attention on what's happening, and increases to that extent the potentials for tension.

Is there a way out? I'm not sure there's any easy answer. There may be a procedural way out, although a difficult one. It is to try to define methods of bargaining.

Thirty years ago this society was faced with a huge number of labor crises which almost pulled it apart. There were sit-down strikes, seizures of property—conflicts on a scale far beyond anything that's happened in the schools in the last ten years. And yet there was a settlement on the basis of bargaining. It doesn't always work, but by and large there is a locus of bargaining. It's helped us arrive at some of the most important decisions this country has faced. Now we may have to devise methods for political bargaining, in which we have on a community basis a higher degree of cohesiveness and coherence of groups and a sense of trade-offs. What do you want, and what are you going to give up for what you get?

THE POST-INDUSTRIAL SOCIETY

I have left almost no time for the third of my topics, which is my topic itself: the post-industrial society. This is, in a sense, the picture of our society thirty years hence. Let me sketch two very broad notions of a post-industrial society.

The national society is a society we've been inheriting for the last thirty years. The communal society is a way of reconceptualizing what's upon us. The post-industrial society is a way of conceptualizing what's ahead.

Most of the world today is still pre-industrial. More than 60 percent of the labor force (in areas such as Asia, Africa and Latin America) is basically engaged in extractive industries like agriculture, fishing and mining. These are the primary industries of those societies. Their technology is basically that of raw materials. Their labor force is unskilled. They are playing a game against nature, because they are earning their living from nature, and this is subject to the law of diminishing returns.

Another band of societies are those of western Europe, the Soviet Union and Japan (the United States is moving out of this category). They are industrial societies in that they are primarily goods-producing; the weight of the economy is in manufacturing or processing. The predominant worker is either the semi-skilled worker learning a job which he can figure out in two or three weeks, or an engineer. Just as the pre-industrial society is a game against nature, the industrial society is a game against fabricated nature. It's a man-machine relationship, in which the man is sometimes an appendage to the machine. But it's based upon productivity and growth, because it increases the efficiency of goods in society, as against diminishing returns.

The post-industrial society is a service society. Today six out of every ten persons in the United States are engaged in services; by 1980 seven out of every ten will be. A service is trade, finance, real estate, insurance, health, education, research, government, transportation, recreation et cetera. These are fundamentally white-collar employments.

The technology of a pre-industrial society is essentially raw materials. The technology of an industrial economy is essentially energy. The technology of the post-industrial society is essentially information. And whereas a pre-industrial society is a game against nature and an industrial society is a game against fabricated nature, the post-industrial society is a game between persons. Its relationships of work are essentially games between people, not against machines or against nature itself.

This is a sketch of a concept which you can lay out on a grid. Then you can spend many hours, as I do in my classes, trying to deal with its consequences.

There is another axis, however, which is even more important in understanding the centrality of the post-industrial society. Here my use of the phrase is very different from that of Herman Kahn, who largely concentrates on the idea of a service economy as the basis of the post-industrial society.

It seems to me that the most fundamental aspect of the post-industrial society is not so much its changeover to services—important as this is—but the centrality of theoretical knowledge. Though every society has been dependent upon knowledge, very rarely has this been theoretical knowledge. It's the codification of theoretical knowledge that now becomes the source of advance and change in the society.

Looking at the relation of science to technology, consider the major industries we still have — steel, auto, aviation, electricity, telephone, telegraph. You find that these are all nineteenth-century industries (although steel begins in the eighteenth century with Darby and the coking process, aviation in the twentieth century with the Wright brothers). But they are centered in the nineteenth century insofar as the pattern of innovation and the pattern of growth are concerned. They were all created by talented tinkerers working independently of the basic laws of science. Darby, who devised the coking process of steel, or Siemans, who created the dynamo, or Edison, who invented the electric lamp, or Bell, who came up with the telephone, or the Wright brothers in aviation—all were tinkerers. The first modern industry is chemistry, because you have to have a theoretical knowledge of the properties of the macromolecules you're manipulating in order to know where you're going.

All the major industries that are arising now are the science-based industries — whether they be computers, electronics, optics, fluidics, or focused on a particular product, such as holograms or lasers. And they emerge out of the codification of theoretical knowledge. The function of theoretical knowledge is primarily to reduce empiricism, to provide an abstract system of symbols, mathematical in nature. From these you can derive a whole series of postulates out of the axioms you've created with the basic theorems you've derived from your codification of the symbols themselves. Not only science and technology, but all areas, are moving toward the codification of theoretical knowledge.

Take, for example, the area of government economic policy. The memoirs and diaries of statesmen thirty or forty years ago clearly show that very few of them knew what to do during the depression. A remarkable book, *Diary and Letters,* by a man named Tom Jones, who was the secretary of the cabinet in England at that period, says simply: "We don't know what to do."

Today there is a vast difference in the nature of theoretical knowledge and economics. Most economists know what they are doing, even though they disagree among themselves. The disagreement is less important than the sense that they have some notion of what to do, even though the politics of it may be difficult.

Recently we have had the extraordinary spectacle of a labor government in England deliberately engineering a recession to redeploy resources. This action was based on Nicholas Kaldor's theory of a selective employment tax.

It's striking to realize that a concept like GNP, which we use so easily and glibly, came into common use only in 1945. It was first proposed in Franklin Roosevelt's budget message in 1944. The notation of GNP originated in the thirties with the work of Simon Kuznitz, and became a government planning mechanism through the creation of the Council of Economic Advisors. Almost all the macro-economic tools we have are only twenty-five or so years old.

In one respect, since the post-industrial society is basically dependent upon the codification of theoretical knowledge, its primary institutions are intellectual bodies: universities, research organizations, industrial laboratories, and the like. But this doesn't mean that most people will be intellectuals. Ours has been a business civilization, and the business firm was a marvelous social invention of the last hundred years to harness men and materials and markets for mass production of goods. It has been a business civilization though the majority of people have not been businessmen. The values of the society have been essentially business values. The highest rewards of status have gone to businessmen. To say that a new society will be essentially intellectual is to say in effect that there will be new primary institutions which will be essentially intellectual in nature.

Here is a simple illustration. The key strategic resource of the post-industrial society, once you begin to think of it in terms of the necessity of the codification of theoretical knowledge, is essentially human capital (talent, brains). A key resource of an industrial society has been money capital. We couldn't have a large-scale industrial economy until we were able to institutionalize the development of credit mechanisms. We had to tap large amounts of savings to generate savings, and to tap them through insurance pools, the stock market, and various other forms of getting equity money, debenture money et cetera. We know now how to raise money capital: we restrict consumption and use the resulting savings for investment.

Human capital is a very different kind of problem. For example, we don't know enough about the genetic distribution of talent. This does not imply a correlation with race, though most people assume it does. It's not necessarily true at all.

We do know that there is a genetic basis for talent, in the sense that people inherit certain abilities. We also know that they inherit not a fixed IQ but a range, just as they inherit a range of heights. Nobody's genetic heritage specifies his height; he inherits a size in which he ranges between, let's say 5'3" and 5'10". And whether he goes to the upper limit of that range depends upon the nutritional patterns of his childhood. Japanese children today are about half a head taller than their fathers. There's been an increase of four inches or so in the Japanese height, simply because of changes in nutritional patterns.

In the same way, a person inherits a range of intelligence. And scientists are beginning to realize that largely half of the ability to use that range—the capacity to learn—is fixed before the age of five, and the other half before the age of fifteen. After that it doesn't mean that you're finished. Then you have to have wisdom, you see. Then you begin to use your learning for various kinds of ends.

The real problem of the post-industrial society is the question, How do you husband human capital? How do you identify talent at an early age? How do you motivate a person? How do you provide a curriculum which is adequate for what goes on in the society? How, in effect, do you use your talent in the most effective way?

This involves a fifteen- to twenty-year planning cycle, which is very different from a planning cycle of economic resources. It is an illustration of one of the kind of problems which must be solved in the post-industrial society.

I've gone far beyond my time. I'm sorry if I've disheartened you by the fact that after the age of fifteen, your intelligence capacity is restricted. As I said before, what happens is that you begin to learn wisdom.

And if you say, how do you become wise? I can only tell you about the rabbi who was asked by a disciple, "How do you become wise?" And the rabbi said, "Well, I suppose you study and work hard." The disciple said, "You know, Rabbi, a lot of people who study and work hard aren't wise." And the rabbi said, "Well, I suppose you study and work hard, and have experience."

"But Rabbi, a lot of people study and work hard and have experience, and still aren't wise."

And the rabbi said impatiently, "Well, you study, work hard, have experience and have good judgment."

"But rabbi, how do you have good judgment?"

The rabbi said, "By having bad experience."

Thank you.

DANIEL BELL

Daniel Bell is a professor of sociology at Harvard University and has worked principally in the areas of political sociology and social change. Born in New York City, he did his undergraduate work at City College and his graduate work at Columbia University. He has had a mixed career in journalism and the academy. From 1940 to 1944 he was first a writer and then managing editor of *The New Leader;* from 1945 to 1948 he taught social science at the University of Chicago; and from 1948 to 1958 he was an editor at *Fortune* magazine. Since 1958 he has been in the academy. In 1958-1959 he was a fellow at the Center for Advanced Studies in the Behavioral Sciences at Palo Alto, California and from 1959 to 1969 he taught at Columbia University. Since 1969 he has been at Harvard.

Professor Bell maintains his editorial interests as co-editor of the quarterly *The Public Interest* and as a member of the editorial boards of *The American Scholar* and *Daedalus.*

His involvement in social policy has led to service in government. In 1965-1966 he was a member of the President's Commission on Technology, Automation and Economic Progress, and in 1966-1968 he was co-chairman of the Panel of Social Indicators in the Department of Health, Education and Welfare. This panel produced the document *Towards a Social Report.*

Dr. Bell is a fellow of the American Academy of Arts and Sciences and its vice president for the social sciences.

BIBLIOGRAPHY

"Marxian Socialism in the United States." In *Socialism and American Life.* Edited by Donald Drew Egbert and Stow Persons. Princeton, N. J.: Princeton University Press, 1952. Paperback, 1967.

ed. *The New American Right.* New York: Criterion Books, 1955.

Work and Its Discontents. Boston: Beacon Press, 1956.

"*The End of Ideology*": *On the Exhaustion of Political Ideas in the Fifties.* New York: Free Press, 1960.

ed. *The Radical Right.* New York: Doubleday & Co., 1964.

The Reforming of General Education. New York: Columbia University Press, 1966. Paperback, New York: Anchor, 1966.

ed. *Toward the Year 2000.* Boston: Houghton Mifflin Company, 1968. Paperback, Boston: Beacon Press, 1969.

and Irving Kristol, eds. *"Confrontation": Rebellion in the Universities.* New York: Basic Books, Inc., 1971.

The Coming of the Post-Industrial Society. New York: Basic Books, Inc., 1972.

3. THE STATE OF OUR ENVIRONMENT

FRANK GRAHAM, JR.
Field Editor
Audubon Magazine

Early last fall I got a telephone call from a captain in the United States Coast Guard in Washington. He asked me to come down to New Orleans to take part in a symposium which President Nixon had asked the Coast Guard to organize in connection with hazardous substances (pollution resulting from careless handling, and so on). He said that they had invited a number of the leaders of industry and government to take part in a series of symposia in order to make recommendations to the Administration for stronger legislation in this field.

I asked him what part I was expected to play, and he said, "Well, you know, these fellows will be sitting around in these symposia all day. They get pretty bored after a while. We want you to come and speak at lunchtime. After they hear somebody from the Audubon Society, they'll be so mad they'll go back to work with a will and we'll finally get this thing out." So that may have been Jim Barr's idea in inviting me to talk to you today.

Just the fact that so many of you have showed up here this afternoon is an indication of the real interest in our environment that has developed in the last few years. This wasn't always the case. As Jim mentioned to you before, five or six years ago I began to do research on a book about water pollution, and most of my friends had no idea what it was all about or what the subject involved. I got a lot of witty remarks. One wag said that since I was writing a book about sewers, it was bound to be an underground best-seller.

At about the same time there was another writer I know who was preparing a book on air pollution. He got a little discouraged. He said he went around, he talked to people about it. Nobody had heard of it. One fellow thought that air pollution had something to do with the static on his radio.

Murray Stein, who is the chief enforcement officer of the Federal Water Quality Administration, told me that back in the 1950s he was sent out to the west coast to deal with a particularly vexing industrial pollution problem. While sitting around a table with some of the industrial leaders there (after Stein had made certain demands to clean up), one of the men from the polluting industry threw down his pencil in anger and said: "Pretty soon we'll be forced to hire a vice president in charge of pollution control." And Stein said: "You remind me of the people back in the 1930s who, after a few strikes, went around saying, 'If things keep on this way, pretty soon we'll have to hire a vice president in charge of labor relations'."

Well, both of these things have come to pass, as you know. And water pollution and air pollution have become a part of our everyday vocabulary. But as people have begun to learn what they are all about, I think that the word ecology has become a stumbling block. "We've got to do something about our ecology," I heard somebody say the other day. As if ecology itself were some kind of dreaded ogre, like cancer, contamination or even the substances that are put into spray cans of insecticides.

But ecology, as most of you know by now, is the science or the organized body of knowledge that deals with interrelationships among living things and their environment. It isn't a brand-new word like so much of the jargon that we hear today. The *Oxford English Dictionary* traces it back to 1873, when it was first used. And it is derived from two Greek words, meaning literally "a study of the house."

Now, if by house we imply the earth, its life and its envelope of air, we have suggested pretty accurately its true meaning. And if ecology is to describe, again accurately, "a true state of our house," it may justly lay claim to being the modern version of the dismal science.

Now, the root of our trouble is that even people who know that ecology is a flourishing young science (and not a Pandora's box of thunders and stinks) often haven't grasped the full meaning of the word. Ecology, if it implies anything at all, implies complexity. We have come to our present plight (and I think it's even more serious than a lot of the public really knows) not through wickedness or indifference, but chiefly through a lack of understanding of the complexity of the world we live in—in other words, the ecology of the world.

So who's to blame? That's the question you always hear.

I'm sure that all of you know about Pogo's remark when speaking of the environment. He said: We have met the enemy and he is us. Well, that's very amusing and there is some truth to it, but I think when this self-accusatory line is accepted as literally true, then we are no better off than we were when we started. The proposition that all are guilty eventually boils down to the proposition that no one is guilty. This isn't a popular conclusion to come to in a day when the tendency is to embrace any cause with a kind of religious fervor. The doctrine of original sin is still strongly entrenched in all of us.

But I'm not going to stand up here and pad out my environmental demonology with the fellow who goes down to the local automobile dealer and buys one of those portable pollution machines we know as an automobile; or the woman who goes to the supermarket and brings home a six-pack of non-returnable bottles.

Government and industry in effect have left us no choice. I'm going to try to be a little bit more selective here today, and maybe even at the risk of sounding blunt at times. I think that some people don't like to get too specific about this, even within the environmental movement itself, because they say we're running down our potential allies. But I would like to throw some things out to you, and we can go into it a little more afterwards in our discussion.

I want to list now my public enemy No. 1. I think it's an old human trait to package things in threes; a sort of trinity affinity, you might say. So I've assembled three sets of villains for us, all of whom have worked diligently, and even with the best of intentions, to bring America to the edge of environmental disaster. And they are probably, in the inverse order of their destructiveness, first, the legislator who wants to do right by everybody; second, the bureaucrat who is defending his own little fiefdom, or whose interests have made him indistinguishable from the people he is supposed to regulate; and third, the public-spirited industrialist. I want to make it clear in each area just what I'm referring to, because I think this has been part of our problem. The legislator, the bureaucrat and many of the best-intentioned industrialists have contributed in no small way to the problem, and perhaps even to the bulk of it, because in each of these cases we have been lulled into a false sense of security.

THE LEGISLATOR

Now, a cynic would say that the legislator is simply trying to win votes by pleasing everybody. The more charitable would really believe what the legislator has to say—that we can have it both ways.

For instance, that we can have fish and oil existing side by side in one bay.

In any case, the solution provided by these legislators to all of our economic problems invariably turns out to be putting people to work destroying the environment.

They want to mix economics with ecology. Not being experts, their economics often turns out to be as cloudy and as muddled as their ecology.

Let's consider for a moment the SST, which I think is a real area of controversy, because it is not one of these black and white areas. As all of you know, there is a serious economic problem in the Northwest, and the easiest way out for legislators in any of these problems always seems to be to turn to the old pork barrel. The Columbia River and other rivers in the Northwest have been tamed, so that what were once wild rushing rivers are now in many cases simply a succession of placid lakes. The legislators have run out of spaces in which to authorize dams that can be built to give an infusion into the local economy. So the local legislators out there have fastened onto the SST.

Now, who is pushing the SST in Congress? Prominent among them, especially in the Senate, is a man who has a very high reputation in this country as a conservationist: Senator Henry Jackson. He has such a high reputation, in fact, that Charles A. Lindbergh, who is probably the most private of public men, recently came out of his ordinary seclusion to pay tribute to Senator Jackson as a conservationist during a major banquet in Washington.

Now, the SST may not be the environmental disaster that some people have painted it. I think some areas of real discussion exist there. But even Senator Jackson must be aware that the SST at best will be a public nuisance.

In operation it will whisk a handful of well-to-do people to Europe a couple of hours faster, perhaps—if you can get them to the airport in time. But for the rest of us it will further degrade just a little, perhaps, the quality of our lives.

Senator Jackson, the conservationist, has weighed that degradation against the economic lift that it may give to Seattle, and he has opted for the degradation. This is certainly bad ecology.

And in view of the fact that economic alternatives have not been fully explored, and the fact that many of the SST's projected benefits, in the opinion of a number of economists, may have been exaggerated, this plan could be bad economics too.

We don't have to go all the way to Seattle to find a similar situation. We had a taste of it not long ago in my own state of Maine. Some of the elected officials there from both parties were instrumental in having the classification of a stream lowered so that a sugar beet plant could be built on it. It was done with the best of intentions, and among the men who did it was the most noted environmental crusader in Congress. It was done with the intention that it would simply be a temporary situation. Then things got out of control. You can't always keep the lid on. The water went from bad to worse. There was one particularly bad summer when the stream (which unfortunately is not confined to Maine but runs across the border into New Brunswick) so infuriated the Canadians that early one morning they came up to the border with three bulldozers, put up a dam and turned the stream back into Maine. This was very embarrassing for the customs officials, the mounted police in Canada and high officials in both governments. It was so embarrassing, in fact, that Senator Muskie finally had to issue a press release from his Washington office trying to explain away his part in the whole affair. But this affair was the result of a bad decision, hastily taken, though argued against by local conservationists. All in all, it was not an irrevocable mistake.

Another project on the coast of Maine about thirty miles from where I live could be even more damaging. The plan is to build a giant oil port and oil refinery at the town of Machiasport, where the tankers could come in. The only reason for going there is that Machiasport has very deep water offshore, and it's one of the few ports in this part of the country that could handle the giant super-tankers which are being built in Japan now.

Now, the county is depressed and they need jobs. But what they are doing is sacrificing jobs. At the present state of our technology, fishing, clam digging, and many other industries are not compatible with large oil developments. At the beginning of this project many of our leading congressmen jumped on the oil bandwagon. I want to read here a statement that Senator Muskie made several years ago in 1968 in favor of an oil port at Machiasport.

He spoke of the controversy, and I quote here, "as a product of competing interests in land and water. Workmen out of jobs are locked in low paying jobs . . . want industrial growth for new opportunity. Summer residents and visitors and suburban residents want to limit industrial growth as a protection against interference in their own life styles."

But I think what many scientists and new conservationists are saying, as loudly and as urgently as they can, is that pollution control is no longer a matter of life styles. It is a matter of survival. It was this misunderstanding, then, that prompted the rush by public figures to leap aboard the Machiasport bandwagon in 1968. I will say this, that Senator Muskie did call a hearing of the subcommittee at Machias this fall, in which he had testimony from a number of leading scientists. They gave evidence about the state of our ability to clean up oil spills, to control oil around these terminals and such. After the hearings he backed down on his original support for the project. He said that at this point he would vote against any such intrusion of an oil refinery into an unspoiled section of the Maine coast.

But too many times the earlier feeling that I was speaking about is the one that's reflected in Congress. Great environmental disasters are to be avoided if possible, but one can with impunity nibble away at the environment, or even at human health, if it is a matter of economics.

In Washington some years ago I talked with a young lawyer who worked with Senator Kefauver during the drive to get the famous Kefauver drug bill through Congress. This man said: "Time after time we were defeated. We had the facts. We had scientific projections of serious harm, serious consequences, if certain conditions existing in the drug industry were not regulated. But we lacked one thing, the one thing we needed to get that bill through: bodies. It's a terrible thing to say, but Congress never would have passed that bill if we hadn't gotten the bodies, the deformed bodies of the children who were affected by the drug thalidomide."

I'd like to point out here that Senator Muskie almost alone among U.S. senators (Gaylord Nelson of Wisconsin is another) now seems to have grasped the extent of our environmental problems. In the Machiasport controversy that I spoke about, there was some graphic testimony by a member of the Woods Hole Oceanographic Institute. It brought out facts which had not been given general circulation before, about the toxic hazards of oil after oil spills, especially in protected areas where it cannot be completely disseminated out in mid-ocean. I think that the optimism many people once felt, their confidence that we had moved ahead to solve many of these problems, was considerably dimmed by this forecast. This is not to say, of course, that oil and water are irrevocably asunder. Certainly as the state of our technology advances we will, perhaps within a few years, solve these problems. But we have too often seen foul-ups like the oil spill at Santa Barbara, or the break-up of the tanker Arrow off the coast of Nova Scotia. This was before all the talk of the new technology of clean-up—and what did they finally use to clean it up? Straw! They put straw on the beaches, and that was the only weapon they had to deal with the problem. And after they burned the straw, they had an air pollution problem. In other words, we are beginning to learn that letting in a little bit of pollution is like being just a little bit pregnant. Once started, the problem is likely to grow. Until the capacity to move, to be shaken up by testimony, becomes general in Congress and in our state legislatures, ecology will always lose out to unsound economics.

THE BUREAUCRAT

Now we approach the bureaucratic mind. Here too we find men and women who usually want to do right, though sometimes one wonders. I've a little item here I clipped from a recent issue of *Conservation News*. It says: "To illustrate the tense environmental concern of today's highway builders, earlier this year Transportation Secretary Volpe issued a glowing press release about the relocation of a Florida highway to avoid an eagle's nest. Alas, we are now disillusioned. It turns out that the nest has been inactive for at least seven years and has been so reported by the authorities. The nest, however, is in a tree on the property of a city commissioner, and the shift in the route means now the highway won't cut through the commissioner's land."

Well, the bureaucratic mind may work in curious ways, but unfortunately it is almost always single-minded. Decisions affecting the environment are made without considering the complexities of the environment. Sometimes the effects are local. Some of you here may remember a few years ago there was an idea to clean up Rockaway Beach with a wonderful new machine called a beach sanitizer. They went in and cleaned up all the garbage that was on the beach, but at the same time the machine also picked up all the shells. So the next wind storm came along and blew all the sand away, and they had to get a bulldozer to push in new sand to remake Rockaway Beach.

But sometimes the problems are more complex, and decisions made in one decade will have far-reaching effects in another. Here's a problem that's a good illustration and that particularly interests me as a field editor of *Audubon*, because it gets into an area that we Audubon Society people do not like to talk about. It's the problem of pest birds.

The gull—what we call a seagull or what ornithologists call the herring gull—was once a real ornament to our coasts. It was not an especially prominent feature of the coastal landscape; in fact, early in this century there was some doubt whether the herring gull would survive at all in this area. The bird was being killed by the gunners of the millinery industry for its feathers, for use in women's hats. People lived on many of the islands which are not inhabited today, and they were great eggers; they'd go out and collect all the gulls' eggs, and the gull became a rapidly disappearing species. So Audubon groups and others pushed through the bill which saved not only the seagull, but also many of the beautiful plume birds like the terns, the herons, egrets and so on.

But the gull began to come back, and then man through his own wondrous methods began to help the gull in an unnatural way, by the formation of the tremendous garbage dumps we see all around. The gull became addicted to the easy life. He went on a dole, so to speak. He found all the food he needed to raise his young. He's a very adaptable bird, and where other birds couldn't make the adaptation, the gull did. The gull soon began to have a population explosion. Boston became a perfect example. Logan airport was built right in the center of an area with three big dumps. There's a dump here, a dump here, a dump here, and right in the center, Logan was put on an island. Now, somebody has said that if a landscape architect had been engaged to design a refuge for gulls, he could not have done a better job than the airport builders did at Logan. They created a big flat space. They put little ponds in the middle. They put a little cover of vegetation, and so on. The gulls would come to the dumps, the three dumps, every morning and eat their fill, and then there was this big beautiful resting area, so they'd all come over and land on the runways. Then when a jet took off, the gulls would be scattered. It seemed funny for a while, but I think there have been more air crashes caused or almost caused at Logan by gulls than by anything else.

So that now our task is to undo the population explosion of this one creature, the gull, which is in fact threatening to wipe out many of the prettier, more desirable species of birds on the New England coast. The Arctic tern and other rare birds are being preyed upon by the gulls, and their numbers are fast decreasing.

So this is an area where mankind created an explosion that we did not anticipate.

In no area of national life, I think, has the bureaucratic mind been less responsive to the complexities of the environment than in chemical pesticides. This is a field I've worked in quite a bit. I'd like to bring up for a moment a little creature called the fire ant. The fire ant is a species that was imported here inadvertently early in this century from South America. It is not a serious pest. In fact, in its home grounds in Argentina, it is coveted by the farmers because it eats a lot of other insects. It has become something of a nuisance in our Southern agricultural fields, because it builds. It's a mound-building ant that builds mounds which are about a foot high, and this often interferes with the operation of farm machinery. It also has a very painful sting, but not nearly as serious as the stings of bees and wasps and other creatures which we think of as being beneficial.

For a number of year the U.S. Department of Agriculture never mentioned the fire ant as a serious pest. But suddenly in about 1956 the Department decided that the fire ant was a major pest to crops and had to be eradicated; not just controlled, but eradicated. They got a tremendous appropriation from Congress. They started, without any previous investigations, a mass spraying program over millions of acres in nine Southern states, which became one of the most notorious environmental campaigns of our time. Destruction was immense. It just killed everything. They sprayed long-lasting pesticides over millions of acres; killed cows and all sorts of wild life: quail, bobwhite quail, woodcock, pheasants, rabbits, alligators.

The thing was fairly well hushed up, even though there was great antagonism within some of the Southern states. Alabama and Florida finally asked to pull out. The story did not really become publicly known until the appearance of Rachel Carson's *Silent Spring* in 1962. The results had been so bad that the Department of Agriculture gradually began to phase it out. In fact, they found that after this tremendous, tremendously expensive, tremendously destructive campaign, at the end of four or five years the fire ant had extended its range by eleven million acres.

Probably no other single event aroused so much indignation among professional people. There had been immense destruction of wild life, forage plants for cattle had been contaminated, the milk itself had been contaminated and condemned. And I think it was certainly one of the big events that spurred President Kennedy in 1963 to ask his science advisory committee to make a full investigation of pesticides. The other event was the publication of *Silent Spring*. In 1963 the committee came out with a recommendation that the persistent pesticides, like DDT, be gradually phased out.

Well, the years have come and gone. The hazards of using many of these pesticides, either carelessly or even scrupulously following the directions, continue to be documented. Yet among certain bureaucrats the notion persists that the best way to deal with nature is to bludgeon it into submission.

Now I'd like to show you a couple of documents here, briefly. This is a little report put out by the government (Committee on Government Operations of the House of Representatives). It's called "Deficiencies in the Administration of the Federal Insecticide Act." The whole thing is a documentation of the fact that the Department of Agriculture, despite the laws that say it should cooperate with other interested agencies, has failed to do so. In 1969 there were 189 occasions on which the Department of Agriculture registered a pesticide or a chemical that had been objected to for health reasons by the Department of Health, Education and Welfare.

A good deal of this report deals with one incident: the registration of something with which many of you are familiar. It is these "no pest" strips which you see hanging in homes, and even though they're against the law, in restaurants and other places where food is prepared. There was a long and involved registration procedure where Health, Education and Welfare had many objections to using the strips, because they give out a continuous vapor. The idea is that the risk day after day in the home is too great. The cost-benefit ratio is really off here. It is the kind of product which the public health authorities do recommend using in certain tropical areas where there is a danger of malarial mosquitos. The danger of malaria, obviously, is greater than the danger of pesticide poisoning from constant exposure to one of these strips. It was developed for tropical mosquito control, but it's being used as a convenience device. After a great deal of fuss, it has now been banned in restaurants, and there is a warning on the label saying that it should not be used in rooms where infants, children, the very old or the infirm are continuously confined.

But anyway, they came to an impasse, and so the Department of

Agriculture appointed a committee to make recommendations as to whether or not this product should be reviewed and accepted. It turned out that two of the men on the committee were employees of the Shell Chemical Company. A third, immediately after he gave his okay to the pest strip, resigned from the Department of Agriculture and joined Shell Chemical.

The report says that the subcommittee has been advised by the Department of Agriculture that possible conflict-of-interest questions involving these men have been or will be referred to the Department of Justice. Nothing has been done, of course, but it has obviously left a bad taste in many people's mouths.

Here is another document of a sort that comes out periodically. In 1963 President Kennedy's Science Advisory Committee advised the phasing out of the so-called persistent pesticides, but every once in a while someone in Washington decides to study it all over again. This document was put out by Secretary Finch's committee—*Report of the Secretary's Committee on Pesticides and their Relationship to Environmental Health*. This tremendous book, probably the most exhaustive which has yet been published on pesticides, documents many things. It comes to the same conclusions that they came to in 1963: that persistent pesticides should be phased out.

In an effort to coordinate, to get away from some of the abuses and disputes between and among government agencies, President Nixon set up in the White House a Council on Environmental Quality. It is to coordinate all these departments so that they are not going off in different directions—to see, for instance, that the Fish and Wildlife Service is not propagating fish while the Department of Agriculture puts DDT in the rivers and kills the fish. Here is the first annual report of the Council on Environmental Quality. It was sent to me by an official of the Department of the Interior. He says: "Dear Frank, I thought you might be interested in the enclosed report. It's a pretty humdrum document, and it's too bad that the Council had to put it out before it really had anything to say but the Act of 1969 required it. From what I can find out here, the Council's being by-passed on many important issues and the agencies are not bothering to report to it. It looks as if it might be a PR setup, at least for the time being."

And what sort of ecological sanity have we come to after all these reports? Well, just recently the USDA has announced the beginning of an enormous program, costing $200 million, to spray by air 120 million acres of land in nine Southern states with 450 million pounds of tiny corncob granules treated with the poison myrex. The object of this is to eradicate the fire ant. Myrex, by the way, according to Secretary Finch's report, is a chlorinated hydrocarbon, a relative of DDT, and is carcinogenic—otherwise known as a cancer-causing substance. The USDA in three successive treatments plans to spray each square foot of land on those 120 million acres of fish, of fields, lakes and rooftops with about fifty of these tiny baits. And remember that this is today and not five years before the publication of *Silent Spring*.

THE INDUSTRIALIST

Well, now we turn to industry. Jim Barr said before: Can you think of any way in which AT&T is an environmental menace? And I'm afraid I haven't been able to figure that one out yet. I guess I can't antagonize anybody here. Someone has complained that telephone books are an environmental menace because there are so many of them—what do you do with them? But this was reported in the *New York Times*—which probably has more paper in its Sunday edition than one of your telephone books.

Obviously, industry is not motivated by conscious wickedness. The devil theory in this area just doesn't work. Businessmen naturally operate on a cost-benefit ratio. They are civic-minded in many ways. But I think that in many cases they have not woven ecological reality into their cost-benefit calculations. For instance, they think of water and air as free, expendable substances, without calculating the cost of their destruction—meaning the cost either of the eventual enormous clean-up or of the effects that this destruction may have on human health and on all kinds of life.

No one can argue that it's impossible for a clean environment to co-exist with industry. It's just that so many of these industries have thought so little about maintaining a clean environment that they've become accustomed to operating in stench and filth. When industry really tries, the results can be amazing. Not long ago I read about a drug company which bragged about the air inside its factory. Its air had to be purified of other spores, so that antibiotics could be manufactured. The firm claimed that the world's purest air was to be found not on some remote mountaintop in the Rockies, but in its own factory. Of course, it didn't say anything about what the air was like outside the factory.

I'd like to look for a moment in this connection at the automobile. Hindsight is always gratifying; it's not always productive. But I think in this case it is very productive and instructive. We all know by now that the automobile is not an unalloyed blessing. In fact, early in this century there was a Boston society matron who is said to have looked over the new horseless carriages running down the streets of Boston. She expressed somewhat the same opinion that we do today. She said that the automobile would divide all mankind into two classes: the quick and the dead.

The people who direct things in Detroit, however, were a lot slower to accept such reservations. Let's recall for a moment the automotive industry's myopia.

Within our own lifetime, since World War II, the automobile became the country's most persistent source of urban pollution: air pollution. Yet the industry did its best to keep this fact a secret. I came across a file of correspondence which was kept by the chairman of the Los Angeles County Board of Supervisors. For a long while he had tried to break the barrier of secrecy. He said years ago that the automobile industry has failed in its responsibility to the American public and to the health of the community. He spoke from considerable experience, because he'd been badgering the industry for so long that his correspondence even includes the Packard automobile among its targets.

He was aware, of course, of Los Angeles's smog problem as early as 1950. The industry paid no attention to his complaints, and he kept sending off letters to the major automakers in Detroit.

Now, here is a portion of a letter that he received on March 3, 1953, from a Ford public relations man. "The Ford engineering staff, although mindful that automobile engines produce exhaust gases, feels these waste vapors are dissipating in the atmosphere quickly, and do not present an air pollution problem. Therefore our research department has not conducted any experimental work aimed at totally eliminating these gases. The fine automotive power plants which modern day engineers design do not smoke. Only aging engines subjected to improper care and maintenance burn oil. To date the need for a device which will more effectively reduce exhaust vapors has not been established."

The Los Angeles man persisted. At intervals he shot off letters to the industry's big three. Invariably the answers were unsatisfactory. A General Motors vice president (at least his correspondents were climbing the corporate ladder) assured him that GM was working hard to reduce pollutants in auto emissions, but the vice president said: "I would be less than candid, however, if I failed to point out that we believe that these changes on automotive vehicles alone will have a very limited influence on the total smog problem in Los Angeles."

The correspondence went on and on, almost to the point of boredom. And what's going on today? Here's a clipping from a recent *New York Times*: "Auto industry changing strategy; open counterattack on environmental and consumer movement."

And so the country's air grew fouler as the people responsible turned their heads. Across the ocean London had not heeded the warnings presented by centuries of smog. Again, people's intentions were the very best. The only trouble was that they hadn't realized the complexity of the environment they were dealing with. They hadn't, in a sense, studied the so-called new science of ecology. In fact, during World War II the British government actually encouraged the production of smog to hide London from the Nazi bombers.

But in December 1952, London paid its bill to nature. The prevailing climatic conditions triggered a killer smog, and the acrid smell of sulphur dioxide permeated every street and every room in the city. It was the odor of hell, and afterward the record showed more than four thousand so-called excess deaths during the few days the city lay in the smog's grip. According to a British medical journal, in the last hundred years only the top week of the influenza epidemic in November 1918 produced more deaths over the expected normal than did the man-made fog. Even the cholera epidemic of 1886 could not equal it.

Yet the most frightening part of the story was that London remained completely unaware of the tragedy it was living through. Radio weather reports gave no indication that the fog was abnormal in any way. Newspapers spoke only of traffic tie-ups or occasional afflicted livestock. Government health authorities issued no warnings. It was days afterwards when the hospitals ran out of beds, morticians began to complain of overwork and the week's vital statistics were recorded, that the smog's devastation revealed itself to its victims.

This, I think, was a grim allegory illustrating the warning that scientists and conservationists have sounded for modern man: that long before we discover the extent to which we have contaminated ourselves and our planet, we may have irrevocably determined our own fate. And it was a warning that millions of Americans began to listen to as the 1970s opened.

To take another industry, and one with which I've had some dealings, the oil industry has fought every meaningful piece of legislation that has been designed to protect the environment. In Maine, for instance, the state is naturally worried about the impact of oil. Oil is not something that's in the distant future; oil is already in Maine. Tremendous quantities are brought in through Portland. More oil comes through Portland on its way to Montreal than is used in the state of Maine itself. Portland is the supply point for Montreal.

So it's there. There have been a number of problems. But the problems will get considerably greater as the new supertankers come into operation.

The state tried to protect itself by setting up some legislation to at least alleviate the damage. One law required industry to pay a tiny fraction of a cent on each barrel of oil going through Maine, to set up a fund which would go to compensate the victims of any large oil spill. Usually these spills are so huge that there is no insurance to cover them. Also, Maine passed a so-called site selection act, which I think is one of the most important pieces of legislation in the environmental field that we've seen in this country. The act says that no large installation of any kind, taking up so many square feet, so many acres, can be built unless it is approved by the state environmental improvement commission. The idea is to head off disaster even before it starts, rather than put, say, a power plant on part of a bay where it may cause very serious damage by either thermal pollution or discharge of toxic metals into the water.

Maine's self-protection was not very strict legislation at all, and it certainly left a place for the oil industry to come in and develop. But instead of living with this, the industry has chosen to take it to court, and is now challenging the validity of the law. Now Maine, which a year ago thought it had adequate protection against the extension of this giant industry, finds itself completely defenseless.

I think all of you have probably followed the automobile industry's moaning and wailing when they were confronted by Senator Muskie's bill to set meaningful emissions standards to protect the public health by 1975.

Not long ago, I watched a television report on mercury contamination, which is certainly a serious problem. I had some reservations myself about the extent of this contamination, but we know that in some parts of the world it has been deadly. It has caused major tragedies in Japan, where factories were pouring mercury into the bay and people ate the fish that were caught in that bay. There were not only many deaths, but mental retardation in children, deformities, that kind of thing. And Sweden had similar unhappy experiences with the use of some methyl mercury compounds. Finally it banned these compounds' use and exported them to the United States, where we now use them to treat wheat seeds.

At the end of the television program, a vice president of the leading chemical firm came on and said that he hoped the government agencies would not do anything rash in making the industry curb this contamination.

In concluding this section on industry, let me just look at the pesticide industry. As I say, it's the one I'm most familiar with through my own work. Its response to the unpleasant revelations today is exactly the same as it was years ago when *Silent Spring* was published. It meets these problems not with technological reform, but with a public relations campaign. There is a difference, however. Its message today is more complex, and therefore more difficult for the layman to see through.

I have a brother-in-law who works for the legal department of a major chemical firm, so we often have it out together. He sends me all this material to try to convert me to his cause, and I use it in lectures.

Here's a publicity piece that's kind of interesting. It was not sent to me by my brother-in-law but by a woman in California. Her husband is an orange grower. It's a copy of the *Sunkist Newsletter*, which is mailed out to all the orange growers in California. It's moaning about the attempts to restrict the use of DDT. And it says: "We were recently handed a treatise prepared by an ecologist (considered one of the nation's top authorities on wildlife biology) earning his spurs before the election year 1970. He spent the first thirty-four years of his professional career with the fish and wildlife service and the last five years as chief staff officer in the pesticide regulation division. Years of service and the title certainly mark him as a man interested in the welfare of wildlife and the possible effect of pesticides on wildlife. Some of his observations should be required reading for thoughtful politicians and amateur wildlife lovers." It goes on to tell some of the things he had to say: that, for instance, DDT is really good for wildlife, and that our streams are not contaminated.

But I want to point out that this was sent out—a glowing testimony to this man—without a mention of the fact that he has long ago left government service and has been for some time the chief spokesman for the National Agricultural Chemicals Association.

Barrons, the national business weekly with which you're all familiar, had a big article not long ago which the Environmental Defense Fund claims was prepared by the same man. The article was called "Ravaged Summer," and it claimed to be a natural sequel to *Silent Spring.* (*Barrons* has been bemoaning the emotional approach of the bird lovers, the mystics, and so on, who want to restrict the unrestricted use of chemical pesticides.) Among the points it makes, tricky points unless you know what you're talking about, is this: "Indeed throughout the U.S., which once could boast of stamping out the anopheles mosquito and its cargo of disease and death, lately have come reports of frightening outbreaks of malaria." Well, that's true; there is a rise in the malaria cases in this country, and in almost every case they have been traced to the veterans returning from Vietnam. There was a similar rise during the early 1950s when the veterans were coming back from Korea. But there is no resurgence of the anopheles mosquito in this country.

Barrons plays on people's politics by mentioning some of the states that have banned DDT, and then saying: "Also banning the use of DDT are the Labor government in Britain and socialist Sweden." Concluding a long diatribe about the emotional approach of the ecologists and conservationists, it says: "By their fruits ye shall know them. Through their unbridled recklessness with facts and sheer irrationality (a triumph of superstition over science) so-called conservationists and ecologists have poisoned the climate of opinion. Now willy-nilly they are threatening to unleash famine and pestilence upon their fellow citizens. They profess to preserve wildlife, defend the environment, befriend the earth. Their natural prey is civilized man."

In answer to an article I wrote for the *Atlantic Monthly,* the scientist whom I mentioned before in connection with the Shell "no pest" strip said in a letter to the editor: "Many of the statements made as proven fact by Mr. Graham are conjecture or open to considerable question. The flat statement that DDT causes cancer in laboratory rodents is wrong. Arguments have raged for fifteen years about this very question. It is far from clear that these tumors are cancerous or that their meaning can be interpreted in terms of man. The Food and Drug

Administration scientists, among others, have not believed these tumors to be cancerous."

Unfortunately for this man, about three days later an article in the *New York Times* stated that DDT had been found to cause mutations in rats. It pointed out that the research, which was conducted by the chief of the FDA cell biology division, was predictable in the light of earlier reports showing that DDT causes cancerous tumors in animals.

The letter to the *Atlantic* editor ended by criticizing my criticism of the "no pest" strips, which I think should earn him some award. I don't want to bore you with all this, but I needed to be specific in showing you some of the things that are prepared by the National Agricultural Chemicals Association. They all tell about the same story: that DDT should not be banned. They say that DDT isn't really the villain in this case; it's something called polychlorinated biphenols, which they claim look like DDT when they show up in these sensitive analyses. Well, it is true that they do appear somewhat the same on gas chromatograph machines, and so on. But any competent biologist is able to distinguish between DDT and the polychlorinated biphenols which occur as wastes in some manufacturing processes.

The Association tries to use the National Audubon's own figures to document the contention that DDT is not hurting the birds but actually helping them, because the counts for starlings, grackles, redwings, cow birds and so on are rising. But the Audubon Society warned them almost ten years ago that these counts were not scientific. In fact, within the society they are known as a kind of bird gulf: the game is to go out and count as many birds as you can. More people go out and count every year. They have better equipment all the time. They're better organized. And, of course, they keep turning up many more birds than they ever did before.

It is true, however, that there are certain species of birds which are certainly rising. They are some of the grain-eating species, which are, for technical-biological reasons, very resistant to DDT. It's usually the flesh-eating species whose reproduction drops. The grain-eating birds are exploding not because DDT is good for them, but because modern farming practices—the giant monoculture, the grain fields, the great cattle-feed lots where tremendous quantities of grain are concentrated—attract these birds. Like the gulls who've been attracted to the dumps, these are the pest birds, the problem birds that are becoming, of course, a problem for the National Audubon Society too.

The other statement which you see in all the Chemical Association's releases is that DDT is not found in serious concentrations in our lakes and streams. That is very true, because DDT is not soluble in water. It is fat-soluble, and studies have time and again proved that when DDT and some of the other persistent pesticides get into lakes, they are absorbed by the living organisms in the water. They adhere naturally to fat, and so a lake like Lake Michigan, for instance, has tiny quantities of DDT in its water and in its bottom sediments, but tremendous amounts in its living organisms.

THE ECOLOGICAL CONSCIENCE

Well, where do we go from here? Ideally, of course, we must work to instill in those who make the big decisions—the legislators, the bureaucrats and the top industrialists in the critical industries that I've been talking about—what one conservationist has called the ecological conscience. Perhaps it will take the revolution to do this— the greening of Ameria which some claim is our only hope of salvation. They may be right, but I do not believe that the revolution is going to take place in this country today or even tomorrow, and in many respects tomorrow may be too late in ecological terms. Not that I see enormous catastrophe looming. I think that the prophets who foresee some sort of global killer smog are really optimists in predicting that a giant cataclysmic event will come along and cause great destruction and nearly kill off the human species.

I think they're optimists because, in a sense, it would be a great cleansing experience to the few survivors who dug their way out of the rubble, and we could start again with a clean slate. No, I see our environment disintegrating not with a bang, but a whimper. I see a gradual degradation . . . a lot of the little things: the loss of a bird here, the destruction of a river or a beach there . . . just a trifle more smog, just a few more sonic booms.

I spoke to a friend whose place I've stayed at sometimes on the Gulf coast of Florida, in the Florida panhandle section. She and her husband have a beautiful house in a magnificent area of dunes. I asked about her place, and she said that people are coming down from other states and bringing dune buggies. They're grinding down the dunes and tearing up the vegetation, and as each wind storm comes it blows off more and more of the sand. She said you can actually see the dunes disappearing.

That's not a big thing, just as the depredations of some of the snowmobilers in our Northern states are not big things, nothing we ordinarily get excited about. But they all play their little part in this story. For the foreseeable future I predict this gradual degradation of our environment. So that in the end, unless we are careful, we will be living in a kind of greyness.

But one cannot act on hopeless assumptions. There is still much that we can do to keep a livable environment, at least in our own lifetime. Perhaps through education—getting people from children up to look at the world around them in complex terms, to observe that things aren't always as simple as they seem, to observe that when you clean up the beach with a beach sanitizer, maybe you'll pick up something more than the garbage on the beach.

I think that through sound laws, through successful court action, the greyness may in part be staved off. The individual can do his share alone or through banding with others in private organizations. But he must have government support. If non-returnable bottles and polluting cars are a menace, manufacture must be stopped, and in one way or another alternatives must be found—must be made generally available, because you can't tell a person not to buy an automobile; you can't tell him not to buy a six-pack of beer, if that's what he wants for his party tonight. There is no reason to put the burden on him when he has no alternative.

I think, also, one of the things which we are coming to realize more and more all the time is that our own reproductive madness has become a menace. I think there are certainly signs in this country that we are coming to grips with the problem. But as you know, in many of the underdeveloped countries for one reason or another, this tremendous population increase is going on, and the end can only be disaster for these people, and perhaps for us.

I remember that Rachel Carson once offered some excellent advice. She said: "Do your homework, mind your English, and care." We ought to remember that. We can at least go down fighting against the expression of another philosophy that's been attributed to the board chairman of General Motors—the immortal words: "Obsolescence is progress." I guess you can say that what the environmental crusade of the 1970s must concern itself with is the attempt to organize people on all levels to keep the pressure on both government and industry. To vote the laggards out of office. In the words of one environmental lawyer, "To sue the bastards." To try to learn enough about what is happening to us so that we can apply political and economic pressures on the polluters and the despoilers.

The history of the new conservation has yet to be written. It should be a splendid and perhaps hair-raising adventure story, and the outcome will be in doubt to the very end.

In this corner, at least, the notion persists that unless the new conservationists (a term which by necessity now includes all of us) are able to make the connection between, on the one hand, all of the wonders of the natural world and, on the other hand, themselves . . . then the story cannot have a happy ending. Man, our instinct for survival tells us, will undoubtedly persist. He will make an accommodation to the great invisible machinery of nature upon which our existence depends, but he will have blighted for his remaining span of time here those tangible, wondrous and fleeting blossoms of creation by which we at present measure our humanity. Thank you.

FRANK GRAHAM, JR.

Frank Graham is one of the country's leading authorities on ecology and the state of our environment, a reputation confirmed by the praise and public enthusiasm that greeted the publication of his book *Since Silent Spring* in 1970. The book describes the successful effort by Rachel Carson to break the information barrier and make the American public aware of world-wide contamination by chemical pesticides.

Mr. Graham began his career as an ecologist and conservationist in the early 1960s with an investigation of strip mining abuses in Pennsylvania. He became increasingly interested in man's contamination of his environment and in 1966 published *Disaster by Default: Politics and Water Pollution,* a landmark study that attracted wide attention among conservationists. The book resulted in a grant from the Rachel Carson Memorial Fund to study recent state and federal legislation concerning the use of pesticides.

Mr. Graham also pursues his concern for ecology as field editor for *Audubon Magazine* and contributor to such magazines as *American Heritage, Sports Illustrated* and *The New Republic.* He was discussed conservation subjects on various television programs, including "The Today Show," "Focus on Books," the "Martha Deane Program" and "Book Beat."

Before Mr. Graham became a champion of conservation and ecology, he was known primarily as a writer about sports, especially baseball. After graduating from Columbia University in 1950, he became publicity director for the Brooklyn Dodgers Baseball Club. Soon he began to write about baseball and in 1956 he joined the staff of *Sport* magazine as assistant manager. In 1958 he resigned to become a free-lance writer. By the 1960s his primary field of interest had become the many threats to our environment. Today Mr. Graham is a member of the National Audubon Society and the Authors League of America, and lectures extensively on the subjects of ecology and the pesticide controversy. With his wife, Ada, he has also written books about nature and conservation for children, including *Wildlife Rescue* and *Puffin Island*.

BIBLIOGRAPHY

"Disaster by Default": Politics and Water Pollution. New York: M. Evans & Co., 1966.

Since Silent Spring. Boston: Houghton Mifflin Co., 1970.

"Man's Dominion": The Story of Conservation in America. New York: M. Evans & Co., 1971.

4. POPULATION AND THE HUMAN ENVIRONMENT

PHILIP HAUSER
Professor of Sociology
University of Chicago

In considering the problem which confronts us when we look at population and environment, I would like to begin in true professorial fashion with a time perspective. Man, or some close kissing cousin of man, has been on this earth for perhaps four million years. In the course of his occupation of the planet there have been four developments which have primarily affected his attitudes, his values, his institutions and his behavior.

These four developments, if I may use more or less popular language and apologize for the neologisms involved, are the population explosion, the population implosion, the population displosion and the accelerated tempo of technological change.

I regard these developments as the four elements of what I call the "social morphological revolution." Since the economists have an industrial revolution and engineers have a technological revolution, and so on, I think you'll grant me that the sociologist ought to have a revolution. His is the social morphological revolution.

Let me quickly define these terms.

Most everybody now understands what is meant by the population explosion. It refers to the dramatic acceleration in the rate of world population growth, particularly during the three centuries of the modern era.

By the population implosion I refer to the increasing concentration of people on relatively small portions of the earth's surface. This is the phenomenon better known as urbanization.

By the population displosion I've taken an archaic word out of the dictionary to refer to the growing heterogeneity of people who share not only the same geographic locale, but increasingly the same life space, that is, social, economic and political activities.

By heterogeneity I am referring to diversity of population by culture, language, religion, value systems, ethnicity and race.

These developments, along with the accelerated tempo of technological change, are interrelated. The population explosion fed the implosion; both fed the displosion. Technological change has generally preceded social change and has been both antecedent and consequent to the other developments.

Let me provide a brief overview of these developments in terms of the global scene, and then proceed to focus on the United States.

GLOBAL SCENE

I begin with the global scene not only because part of my assignment is to provide a mind-stretching presentation, but also because I think an understanding of worldwide developments is essential to our purposes. It is a prerequisite to understanding the climate in which American business, including AT&T, will conduct its affairs for at least the remainder of this century.

Explosion

Now the population explosion for the world can be summarized very quickly. It took the predominant proportion of the four million years that man has been on this earth to generate a population of one billion people. That number was not achieved until approximately 1850. But it required only eighty years to add a second billion. Two billion was the world population in 1930. It took only thirty years to add a third billion. Three billion was the world population in 1960.

Should the present rate of growth continue with present birth rates and trends in mortality, world population by the end of the century could be 7.5 billion. Allowing for appreciable reductions in fertility, particularly in the developing regions of the world, in Asia, Latin America and Africa, world population in my judgment could easily approximate 7 billion by 2000. That includes everything I think can be accomplished in reduction of birth rates through family planning programs now underway in some of the more populated areas of the developing regions.

That, in a nutshell, is the story of the population explosion.

Implosion

The population implosion, of course, refers to urbanization. In quick overview let me remind you that although man has been on the earth for perhaps four million years, he did not achieve permanent settlement until the Neolithic period, about ten thousand years ago. The development of cities or towns, something beyond the Neolithic village, required considerable amounts of time. The calendar for world urbanization indicates that we did not begin to get population agglomerations that could be called towns or cities until between 5000 and 3500 B.C.

Now, to achieve aggregative living requires, among other things, two fundamental types of development: technology and social organization. Mankind as a whole had not achieved enough in technological development on the one hand and social organizational development on the other to create cities as large as 100,000 until as recently as Greco-Roman civilization. Mankind did not achieve enough development, technological and social, to permit cities of a million or more until as recently as 1800. To be sure, there is some debate about whether ancient China had a city of a million or more. The more research that is done, the less likely it seems. Probably the first city of a million or more was the precursor of Tokyo: Edo. Paris and London have claimed that distinction, but they cannot prove it.

If we consider world urbanization, the population implosion, it is clearly a phenomenon much more recent than the population explosion. If we use the definition proposed by the United Nations for purposes of international comparability (that urban places be deemed places of 200,000 or more persons), then as recently as 1800 only a little more than 2 percent of the world's peoples lived in urban places. By 1950 about a fifth of the world population, about 20 percent, was urban. At the present time it is probably around 28 percent.

But consider this fact. Should present trends in world urbanization continue, it is possible that by the end of this century, which is a little more than one human generation away, almost half of the world's people will live in urban places—48 percent, according to the projections of the United Nations. Urbanization represents a tremendous transformation, obviously, in the way of life for the world's peoples. From about 2 percent to half the world's population urbanized between 1000 and 2000 is the consequence of the population implosion.

Displosion

Now, let us consider the population displosion on a global basis. This is a little harder to quantify. Let me just say that as travel and communication were greatly improved, largely through technological advance—an advance in which your company certainly has played an outstanding role—the world has shrunk. A major consequence of the shrinking world has been tremendously increased interaction among diverse peoples—diverse by culture, language, religion, values, race, ethnicity and so on.

This increased interaction of diverse peoples, as in the case of the other two population developments, has generated many problems which afflict the entire world today, including the United States. One of the major consequences of the population displosion is the conflict that is evident throughout the world among diverse peoples still trying to learn to live together, not only in the same geographic area, but also in the same life space.

Moreover, the population displosion, in the sense in which I'm speaking of it, is as recent as the end of World War II. Only since then, and for the first time in human history, by reason of what Adlai Stevenson popularized as "the revolution of rising expectations," has it been true that there are virtually no peoples left on the face of the earth who are willing to settle for second place in anything, and who do not insist upon freedom and independence if they have not yet achieved them.

What are the manifestations of the conflict to which I have referred? Examples are the conflicts between Protestants and Catholics in Northern Ireland; between Israelis and Arabs in the Near East; between the Malays and the Chinese in Malaysia; between Hindu and Moslem in India and Pakistan; among the diverse linguistic groups of Hindus within India; among the tribal societies in Africa; between black and white in the Union of South Africa, Rhodesia and the United States of America. All of these illustrate the impact of the population displosion.

Now, in ancient times, to be sure, there were peoples of diverse backgrounds sharing the same geographical locales, as in ancient Rome. However, in ancient societies, and in societies in general before World War II, there was little conflict because they were highly stratified. The non-Romans in Rome were slaves. And relatively speaking, there was little conflict in the United States between black and white under the condition of slavery, where each element of the population "knew its place." But friction was generated by the revolution of rising expectations since the end of World War II, with the expectation of egalitarianism and complete access to the economy and society on the part of all minority groups.

Global Problems

Each of these elements of the social morphological revolution precipitated a whole series of unprecedented problems. Let me just quickly enumerate some major issues. The population explosion, of course, has as its most critical consequence the inability of the developing nations in Asia, Latin America and Africa to achieve their national aspirations for higher levels of living. Excessive population growth precludes the achievement of great increases in productivity and therefore of increases in per capita income.

Basic problems have been raised on the global scene concerning the relation between population and food supply. And here may I say that the literature is quite confusing. You often hear the same author cited as predicting that there will be mass starvation during the seventies and eighties in the developing regions of the world and as predicting that there will be no starvation for several decades to come. A good way to unscramble that conflict in the literature is to get the date of publication. In general, if what you were reading was published before 1967, the outlook was pretty dismal. In the first half of the sixties, population growth was outrunning the rate of food production in Latin America and in Asia and was threatening to do the same in Africa.

But with the so-called "green revolution" since 1967, all predictions about dire consequences in the form of mass starvation, in which tens of millions of people would die in the developing regions, are no longer in order. On the contrary, many areas which had food shortages in the early part of the sixties now have food surpluses and are exporting food. The Philippines, where I was just a few weeks ago, is a good case in point. This does not mean that the green revolution will solve the food problem.

On the food problem, let me make the following observations. It is a grave mistake and simply inconsistent with the facts to say that mankind is threatened with food shortages in the next two or three decades. It is also a great misstatement to say that the food problem is solved.

By reason of the green revolution, if all works out well—there are some ifs involved—mankind at the most may have gained two or three decades, perhaps four decades, during which there ought not to be any worse relationship between population and food than what we have been experiencing up to this point. So we have bought some time—time during which, presumably, we will try to control population growth rates. There is still a danger that population growth will outrun the food supply in the long run.

May I say on that score, without time to elaborate, that there is no reason yet for us to feel very optimistic about the ability of the developing nations in Asia, Latin America and Africa to control their birth rates. Let me remind you that India has had a form of birth control program as a matter of national policy for some twenty years. And although she did not have much in the way of input until about five years ago, she has yet to produce a measurable decrease in her birth rate. This is not an easy matter. In fact, two propositions help to account for the situation, and I think history would make these propositions stand up. First, there has never been an example of a people who, having acquired education, having proceeded to a higher level of living and having broken away from a traditional order, did not reduce their birth rate. Unfortunately, the converse of that proposition is also true. We have yet to see the first example of a nation mired in illiteracy, poverty and a traditional type of society that has managed to decrease its birth rate. This is something for the future to determine.

What is the world outlook? Well, by reason of these very sketchy facts that I presented, I wind up with a picture which I hope does not discourage you unduly. For at least the remainder of this century I think it is pretty clear that because of the problems generated by the population explosion, implosion and displosion, there will be greater, not less, social unrest.

There will be more, not fewer, political instabilities. There will be greater, not smaller, threats to world peace.

Although the world as a whole is now expending something like 200 billion dollars a year for the military and for armaments, the prospect is that there will be more, not less, money spent for military purposes during the remainder of the century.

This means, if I may translate it into the milieu in which American business, including AT&T, will be conducting its affairs in the rest of the century, that we will have higher, not lower, taxes; that there will be bigger, not smaller, government in the United States. And this prediction is based on international considerations alone. I think it is the global prospect with which the United States will have to cope for at least the remainder of the century. Obviously many questions can be raised about how we cope with this kind of world situation. But I cannot do any more than state it as a problem at this time.

DOMESTIC SCENE

Let me present my basic thesis before I turn to the consideration of these developments in the United States. I have said nothing about technological development, but I think it is unnecessary to do so to almost any American audience, and certainly not to this one.

My thesis is this: First, that man, as the only complex culture-building animal on the face of this globe, has in the course of his culture-building developed a twentieth-century demographic and technological world in which he is still trying to learn how to live--and thus far, not too successfully.

Second, the chaos which has arisen out of these developments—a chaos which is manifest throughout the world, including the United States—is likely to get considerably worse before it gets any better.

Third, the fundamental reason why the situation will get worse before it gets better is that mankind as a whole, including those of us in the United States, is attempting to meet twentieth-century problems with nineteenth-century, eighteenth-century and even earlier value systems, ideologies, forms of government and procedures of government.

Let me try to document my thesis in terms of our situation in this country. But first of all permit me to review these developments in the United States.

Explosion

Consider the population explosion. Let me remind you that when the first census of the United States was taken in 1790, this was a nation of fewer than 4 million souls. By the nineteenth census of the United States, taken as of April 1, 1970, there were 205 million. The prospect for the remainder of this century, utilizing the most recently revised projections from the Bureau of the Census, is that in the less than thirty years between now and the end of the century, a little more than one human generation, we are almost certain to add 75 to 95 million people.

It is still quite possible that we shall have become a nation of 300 million by the end of the century. And this despite the fact that there is much excitement these days about ZPG (zero population growth), and despite the fact that in a White House Report written by the National Goals Research staff, there is a chapter on population which makes the astonishingly illiterate demographic statement that we might achieve a stationary population within a decade.

It is quite likely that before the end of this decade the number of births per year in the United States will top five million and possibly reach five and a half million; more than we have ever experienced before in our history. Why?

Well, this would be true even if our birth rates for women at each age go down somewhat, let alone stay at the present level. And the reason is a very easy one to come by. Most of the childbearing in our society is now done by women twenty to twenty-nine years of age. We are almost one hundred percent a birth control society. Only 4 or 5 percent of the people report that they never used any means of limiting family size and insist they never will use such means. The number of women who are in the child-bearing age group is increasing by 35 percent now and will increase some more beyond 1975.

Why?

Well, these are the products of the first post-war baby boom (1946 to 1957) now reaching reproductive age. In fact our second post-war boom (from 1968 on), the echo effect of the first, would have begun somewhat sooner were it not for the fact that our post-war babies have been experiencing economic difficulties. Their problems have led to an increase in age at marriage and in age at birth of first child.

Contrast the experience of our post-war babies and our depression babies—those born during the thirties. The smartest Americans we ever had were the kids that contrived to get themselves born then. They were smart because they were in short supply when the marriage rate and the birth rate slumped along with the stock market. Because they were in short supply, the youngsters born during the depression thirties had utterly different careers and outlooks.

They were able to get jobs at pretty good wage rates even before marriage. They left home before marriage, lived in their own apartments, and constituted an important element in the demand for housing even before marriage. Throughout their careers they were less subject to unemployment and had more opportunity for promotion than any cohort of Americans we ever had. They experienced a great decrease in age at marriage and a decrease in age at birth of first child. They were also able to have more children per couple because they had it made.

But now the relatively stupid kids that got themselves born as part of the post-war baby boom are in large supply; they have trouble getting employment. Youth unemployment levels are tremendously high.

For the first time in several decades, age at marriage has turned up. And the net effect of this is that the second post-war baby boom has been deferred; but the inevitable has begun to occur.

Let us turn to another consideration that has some bread-and-butter implications for AT&T. The National Goals Research Report has been silly enough to say we might get zero growth within a decade. Of course, some of the so-called "angry ecologists" have been talking about this and making dire recommendations about government controlling the number of births: perhaps having a generation in which no couple is permitted to have more than one child. That requirement would make, I suspect, a large proportion of this audience criminal, if having more than one child were so defined.

Paul Ehrlich has recommended that the mother of the year be the girl who announces she has just had a tubectomy and has adopted two children.

The demographer has an analytical device called the "net reproduction rate." A net reproduction rate of 100 means that the population would replace itself each generation. Translated into children per family, it would mean that each couple, with the present birth rate and death rate at each age, would average 2.11 children. That would be exact replacement, allowing for mortality, and would give us a zero-rated growth in the longer run, as I shall point out in a moment.

Now, at the present time our net reproduction rate is something like 125, which means that in the United States if the birth rate and death rate at each age were maintained, we would produce an increase in population above replacement by 25 percent per generation.

If the net reproduction rate were 75, as it was in urban America during the depression thirties, and if the birth rate and death rate at each age persisted, it would mean that the United States would have been decreasing by 25 percent per generation.

Here's the point: After a net reproduction rate of unity is achieved, it will take seventy years before zero growth is actually achieved.

Why seventy years? It will take seventy years before the age structure of the population reflects the age-specific birth rates and the age-specific death rates that would produce zero growth. This is so because the age structure today is not the product of those birth rates and death rates that would give us 2.11 children per couple, or zero growth. The age structure today is the result of birth rates and death rates at each age limit that go back as far as a hundred years.

What this adds up to for AT&T, and particularly for those who are planning, shall we say, capital outlays looking to the future, is that the American business community is, at least in my judgment, eighty years away from a zero rate of growth. And in all probability more than eighty years away from a zero rate of growth, because I doubt that we will achieve a net reproduction rate of unity by 1980.

Realistically, then, I hope that no one in this company makes the mistake of assuming, if he reads the White House Report, that a zero rate of growth will come by 1980, as a result of which telephone service throughout the United States might stay at the same level as it seems to be in New York City at the present time. You are not faced with any threat of zero growth for a long time to come.

There is therefore plenty of time to get adjusted to zero growth. But let me make one more point. Up until now we have been assuming in this country, and pretty much throughout the world, that more people mean expanding markets, opportunity for more investment and greater return, and increased productivity. Only recently have some people begun to question whether or not we have in many cases passed the point of increasing returns, economies of scale, and arrived at the point of diseconomies of scale. New York City, it is often pointed out, may be one of the examples of diseconomies of scale.

Moreover, it has also been assumed—by Professor Hanson at Harvard during the depression thirties and Sir Roy Harrod in England, also in the thirties—that a zero rate of population growth would mean declining markets and decreased opportunity for business. The interesting thing is that economists are beginning to reexamine this proposition. They are beginning to make a pretty good case for the notion that even with zero growth, a nation can continue to have higher levels of living and expanding markets through increased productivity and increased consumption per capita. There is a whole area to work with here.

Implosion

Let me now get to the population implosion in the United States. When our first census was taken in 1790, 95 percent of the American people lived in rural places. This means places of 2500 or less, as we define them in the census of the United States (not the 20,000 cutoff point between urban and rural, as proposed by the United Nations). There were only twenty-four urban places in the country at the time. Only two of them, New York and Philadelphia, had populations in excess of 25,000. That was the agrarian setting in which the Constitution of the United States was written and in which much of our ideology and value system evolved. By the nineteenth decennial census, drawing on the 1970 census results, we had become 74 percent urban. We are now approximately 69 percent *metropolitan*—69 percent of our population lives in places of 50,000 or more and in the counties where those places are located.

Mark this, and I cannot emphasize the next fact too much—I think it is a prerequisite for understanding what we call our urban crisis. The United States did not become an urban nation, in the sense that more than half of our people lived in urban places, until as recently as 1920. Thus 1970 marked the completion of the first half century in our experience as an urban nation, and a half century is a very small period of time in the life of a nation. And as I like to point out, there are a number of us in this room who know it is less than one lifetime, and hope it is a small part of one lifetime.

Do you realize what that means if you translate it to a micro-personal and familial level? It means that probably a majority of the people in the Congress of the United States were born into an agrarian America with ideologies and values adequately and superbly equipped for agrarian living. It means that the management of much of American business, including AT&T, was born into agrarian America. Congress and the executives in the business world are still trying to make the adjustment from an agrarian society to the urban society that we have become in less than one lifetime.

Let me point out that the increase in the urban population of the United States over the first sixty years of this century absorbed 92 percent of the total growth of the country. Between 1950 and 1960, the increase in the urban population was greater than the increase in the total population. And that is not a mistake in arithmetic on my part. What it signifies is that for the first time in the history of the country, our rural population actually decreased in absolute number.

If we look ahead to the end of the century it is a very safe bet that the total population increment, and perhaps more than the total increment, will become urban between now and the end of the century. Our rural population may well continue to diminish. We could lose half of the rural farm production we now have without decreasing farm production by much more than 5 percent, because there has been such a tremendous increase in productivity. The number of farms in this country has greatly diminished since we first counted them in 1910, but not the total acreage under production. We have much larger farm units with greater capital investment in each unit and tremendously increased productivity.

Displosion

Let us turn next to the displosion. In the United States as recently as 1900, only 51 percent of the people were native whites or of native white parentage. The remainder were foreign-born, children of the foreign-born, blacks or members of other races. As recently as 1960 almost a third of the American people (30 percent) were still foreign-born, children of the foreign-born, blacks or members of other races. We are a polyglot nation. We, as well as the other places in the world to which I referred, are still trying to learn how to live together with diversity in an egalitarian society. And I do not have to spell out the fact that up to this point we have not been too successful in the process, any more than other parts of the world have been.

Domestic Problems

Now these developments in our country have precipitated many problems. Some are problems of the natural environment—environmental degradation, air pollution, water pollution, and the rest. There are some significant problems, also, of the man-made environment—slums, problems in the circulation of goods and persons on the surface, in the air and on water. Some are economic problems—poverty, the revolt of the consumer, unemployment and so on. There are personal and social problems—the revolt of the blacks, the revolt of other minority groups, the revolt of women. I don't have time to elaborate on this last, but *Fortune* (February 1971) quotes me as saying that women are now, under the impact of these developments, in process of change from females to human beings. And, of course, the feminist movement includes those who think the transition is too slow.

There is also the revolt of youth, characterized at one extreme by youth who cannot cope, who seek forms of hippie escape, and at the other extreme by youth who cannot cope and become frenetic actionists, thinking to solve our problems by putting bombs in the men's room in the Capitol. There are also problems of governance—conflicts between federal, state and local government, and the problems of the Congress that cannot get its work done.

Cultural Lag

All of these problems have been generated or exacerbated by the developments to which I have referred. Let me go on in another direction to document an element of my thesis—that we are attempting to deal with our problems, many of which are encapsulated in what we call the urban crisis, with nineteenth-century, eighteenth-century and even earlier value systems, ideologies and institutions. Therefore we shall probably see our society grow more, rather than less, chaotic in the decades which lie ahead.

One of my former professors, William F. Ogburn, introduced into the social science literature the concept of "cultural lag." He perceived that the different elements of our culture change at different tempos, and some lag behind others. Let me give you a couple of concrete examples and then go on to suggest some things that not only may be mind-stretching but undoubtedly will also create some antagonisms.

An example of cultural lag is the provision in the Constitution that it is the right of every American to bear arms. The debates at the Constitutional Convention centered on the right of states to have militia. But this provision has been interpreted to mean that every American has the right to carry a gun.

Let us go back to 1790. The provision made considerable sense then, when 95 percent of our people were living on farms or in small towns. The gun was an important device for increasing the food supply. It was an important device for protecting a family on the frontier from beasts, some of whom were human. And I cannot help but point out the gun was also an important instrument for improving one's realty holdings if Indians were in the way.

I think it is clear that it made considerable sense for every American to have the right to carry a gun in 1790. But I submit to you that to have this right in the last third of the twentieth century, when the nation is 74 percent urban and 69 percent metropolitan, is a fine example of cultural lag. That is why in 1970 there were some 12,000 Americans killed by guns; whereas in comparable-sized populations of about 200 million in Europe or Asia, the number of persons killed each year by guns is to be measured in tens, not thousands.

Or let me give you another example of considerably more significance. I referred to the tremendous rapidity with which we have become urbanized. Well, as recently as 1960 there were 39 states in the Union in which the urban population constituted a majority of the people, but not a single state in which the urban population controlled the state legislature. Not one. The upstate-downstate struggle, with which you are familiar in New York, is seen in every state in which the urban population has become the majority.

Why is the federal government involved in such things as urban renewal, public housing, expressways and highways, civil rights, education, mass transportation and now health? There are some who say it is because the federal government usurps states' rights. I think if you read history intelligently you will come to a different conclusion. It is because state legislatures with dwindling rural minorities have demonstrated perhaps the most harmful form of civil disobedience the United States has ever experienced. In these days when it is customary to speak about civil disobedience, nothing has been as injurious to the American people as the civil disobedience of the state legislatures. For the first sixty years of this century they refused to reapportion and so kept the control of state affairs in the hands of dwindling rural minorities.

These mal-apportioned state legislatures so callously ignored urban problems that they forced the preponderant majority of the American people—70 percent in 1960—to turn to the federal government for the resolution of their problems.

It is not that the federal government usurps states' rights; it is that the state legislatures have defaulted in their obligations to the people. And the net effect is that the state governments have made themselves the fifth wheel of American government.

The effort by the present administration to right the wrong through revenue sharing and the so-called "new federalism" is another fine example of cultural lag. But the man in the White House is not the only gentleman in a position of power in this country with a razor-sharp nineteenth-century mind. Let me give you my interpretation of the proposed revenue-sharing and the new federalism. They are a way of rewarding state governments, which for the first seventy years of this century have refused to allocate resources for basic urban services, including education, public transportation, housing and health. State governments have neither the expertise nor the personnel to deal with urban problems. They are much more subject to special-interest pressures than the federal government—which is not to say that the federal government is not subject to pressure. And they are much more corrupt than the federal government—which is not to say that the federal government is immune to corruption. But coming from the state of Illinois, I can document this with some feeling. Reportedly, Senator Adlai Stevenson said that he dared not fill the shoe boxes of Paul Powell. Powell was our secretary of state who, when he died of a heart attack a while ago, left some $800,000 in currency in shoe boxes in his hotel room.

The new federalism, I would argue, is a feeble effort to go back to where it is no longer possible to go. The trend toward increased concentration of power in Washington is irreversible, as I am sure this administration will discover.

Now, what are some other examples of cultural lag preventing us from dealing with our twentieth-century problems? I shall wind up in a hurry by answering that question and one other: What are some of the major half-truths and distortions about the relation between populations and environment?

We keep repeating what I regard as nonsense shibboleths, which made much sense in their time but are completely inapplicable to the contemporary scene. I apologize in advance if I offend anybody's feelings.

One of them, and I shall give only a few examples, is "That government is best which governs least." Another is "Each man in pursuing his own interests is guided by an invisible hand, so as to act in the interest of the collectivity." Still others are *Caveat emptor*—Let the buyer beware." "Taxes are something government takes away from people and should therefore be kept to a minimum." "A welfare society is a pejorative concept, and must be avoided because we want to remain a rugged individualistic society."

Take the first two. In 1790 it certainly made considerable sense to say that government is best which governs least, and that each man in pursuing his own interest acts in the interests of the collectivity. Then, 95 percent of the people lived on farms or in small towns. What was there for government to do? If you took care of your family on your own farm, you were doing all that was necessary, not only for the economic welfare of your family but also for the economy of the United States. However, we are repeating these shibboleths even though they are utter nonsense in the last third of the twentieth century. Can you imagine the United States without a Social Security system? Without a Pure Food and Drug Administration? (I suppose you could do without an FCC.) We keep repeating them even though they are completely incompatible with the reality of the world in which we live.

I was in a so-called debate with the conservative columnist Kilpatrick—he publishes in about a hundred newspapers. It was at a meeting of the American Home Economics Association in Boston several years ago. Kilpatrick was arguing on behalf of *caveat emptor.* He felt that "Let the buyer beware" ought to be the principle by which the market is regulated in the United States.

But let me interpret *caveat emptor* as I understand it. It means that every woman in this country has the god-given right to have one, two, maybe three deformed thalidomide children before she discovers that it was thalidomide which deformed her babies. Then she could punish the pharmaceutical house by no longer buying any more thalidomide.

Or every housewife has the right, and more of them should exercise it, to learn how to use a slide rule—it can be learned easily. Then when she goes to the supermarket, she can quickly ascertain that if she buys the large-size economy package, it will cost her more per ounce than the small size.

Caveat emptor? Well, this made considerable sense in 1790 when you exchanged a few chickens, maybe, for some flour; but in the last third of the twentieth century, with the complexity of modern products, to say *"Caveat emptor"* is a fine example of cultural lag.

"Taxes are what government takes away from people?" Taxes are funds that pay for services which only government can perform. Now I am not arguing that there are not ways to economize, or that government always performs efficiently. This is a separate issue. But how else are we going to handle air pollution and water pollution and environmental degradation and adequate education? Taxes are something we pay for essential services, and if we don't pay enough taxes, we may be jeopardizing the viability of American society. We may go down the drain of history, as other countries have done, screaming that we kept our taxes low and remained an individualistic, rugged society.

If we cannot afford to pay for preventing environmental degradation, if we cannot afford adequate investment in human beings so that we can decrease the welfare load—we shall be penny-wise and pound-foolish. To my mind one of the most ridiculous spectacles in American history was the sight of the Administration recommending a family assistance program (which had much in its favor, though it contained punitive work provisions) while vetoing appropriations for education and for job training. These contradictory policies would guarantee that the welfare rolls in the next generation will be at least as large as in this.

We are not going to get rid of the welfare load by tinkering with the welfare system. Most of those now on welfare are not people who have fallen by the wayside, but people who never had the opportunity to get underway. And until they obtain an adequate education, until they can command an income flow that enables them to stand on their own feet, we shall have welfare with us.

"A welfare society is a pejorative concept"—a nasty word, socialistic? In our interdependent and highly vulnerable society we cannot help but be a welfare society. Does anyone really feel we can go back again to the days of the thirties, when 25 percent of our labor force was unemployed, when men were forced to sell apples in the streets? You could not get the American people to accept such a situation today. But we keep repeating this utter nonsense. Certainly we are a welfare society. We have no alternative to being one. Among the ridiculous things I have heard in recent years was our President explaining, as he introduced his family assistance program, that this was not going to make us into a welfare society. It was another example of a nineteenth-century shibboleth completely unrelated to realities.

There are other things I could get into that would make you madder than you are now, but I close with a few references to the relation between population and the environment.

Ecology: Truths and Half-Truths

I want to give due credit to the "angry ecologists" who have helped make the American public and the world aware of the fact that there is a problem of environmental degradation and a problem of excessive population growth. But to indicate that there is a one-to-one relationship between population growth and environmental degradation is sheer nonsense. I am delighted to have the opportunity to prick the bubble of the angry ecologists at every opportunity, for this reason: although they are to be commended for arousing public zeal about the problem, I am very much afraid that their distortions, half truths and false predictions may turn public zeal into public apathy when the public discovers it has been "taken." The situation is bad enough without distorting and overstating it.

Here are some examples of the half truths and overstatements. "The population explosion has produced environmental pollution." Utter nonsense. There is no one-to-one relationship; there are many intervening variables. Even if we had zero growth beginning tomorrow, we would still have greatly increased pollution through increased consumption. We could have continuing population growth even while decreasing pollution. "Technology causes pollution." Utter nonsense. Technology, even as science, is a neutral agent. You can have a hammer and make a fine piece of furniture; you can have a hammer and use it to crack a man's skull. "Technology produces pollution" is a half truth; therefore an untruth. Technology can also rid us of pollution.

"A free-enterprise capitalistic society produces pollution." Implicit in this statement is the assumption that a socialist society will not exploit resources and produce pollution. Utter nonsense.

I have spent considerable time behind the Iron Curtain and can report that socialist countries are as polluted as we are, and for the same reason.

Let me put it this way: In a socialist society there is socialized production and privatized consumption on a relatively low level. That society has pollution because it has socialized and deferred paying the cost represented by pollution.

A capitalist society has privatized production and privatized consumption on a much higher level. But it, too, has socialized and deferred paying part of the cost of production. It also has proceeded on the assumption that air was free, water was free, the environment in general was free, and it is just beginning to discover "t'ain't so."

Ken Boulding has stated this better than anyone else, and more succinctly. We as a society, and other human societies thus far, have been prepared to pay for the goods but not for the "bads." And we are just now discovering we must also pay for the bads. We will be rid of pollution only if and when we achieve consensus on wanting to live in a non-polluted environment, and if and when we also agree to pay the cost of it. It will not come free.

The utilities, in my judgment, may be in a better position than any other sector of our economy to get rid of pollution. If the government steps in and sets anti-pollution standards, the utilities can go to their regulatory agencies and say, "The government requires that we increase only costs; therefore we must have an increase in rates." And until government does set standards, it would not make sense for individual utilities to greatly increase production costs unless all utilities are required to do the same.

Let me close with this thought, that the prospect is not necessarily all bleak, though it is the bleak part which I have emphasized tonight. When Clark Kerr, former president of the University of California was fired by Governor Reagan, he made an observation that I think is applicable here. He said as he left: "I leave as I came — fired with enthusiasm."

Philip M. Hauser

Philip M. Hauser, a recognized expert on international demographic problems, is professor of sociology and director of the Population Research Center at the University of Chicago. He and his colleagues at the center are concerned internationally with the economic, political and social implications of the population explosion, including its effects on the world's food supply. Nationally, their major interests include the white middle-class flight to the suburbs which is increasingly leaving the cities to the minority poor and to racial tensions.

Professor Hauser studied at the University of Chicago, where he received his B.A. in 1929, his M.A. in 1933, and his Ph.D. in 1938. He became an instructor there in 1932. While working on his Ph.D. he was associated with the Federal Emergency Relief Administration and the Works Progress Administration. Upon completion of his doctorate he left Chicago to work in Washington for the Bureau of the Census. In 1942 he was promoted to assistant director of the Bureau and four years later became deputy director. While with the Bureau he acquired a reputation as a leading authority on demography and population statistics. In 1947 he rejoined the sociology faculty of the University of Chicago as a professor, but returned to Washington the following year as acting director of the Census Bureau to prepare the 1950 census. He resigned in March 1950, but continued to serve as a consultant.

From 1947 to 1951 Professor Hauser also served as the United States representative to the Population Commission of the United Nations Economic and Social Council. He spent a year in Rangoon, Burma, as a United Nations statistical advisor to help the newly independent government establish a statistical system and plan its first census.

He returned from Burma in 1952 to the faculty at Chicago and from 1956 to 1965 served as chairman of the department of sociology. He helped to reestablish the university as one of the leading centers of sociological graduate training and research, and played a major role in founding the Population Research Center.

During his career Professor Hauser has also been a visiting lecturer or professor at several universities, including Princeton and the Universities of Washington and Indiana, and has served on various government and private committees concerned with population and vital statistics. He has been vitally interested in civil rights in Chicago and was chairman of the Advisory Panel on Integration in Chicago Public Schools.

BIBLIOGRAPHY

and Herbert Blumer. *Movies, Delinquency and Crime.* New York: The Macmillan Company, 1933.

Workers on Relief in the United States, March 1935. Vol. I of *A Census of Usual Occupations.* Washington, D.C.: Government Printing Office, 1938.

and Bruce L. Jenkinson. *Workers on Relief in the United States, March 1935.* Vol. II of *A Study of Industrial and Educational Backgrounds.* Washington, D.C.: Government Printing Office, 1939.

and William R. Leonard, eds. *Government Statistics for Business Use.* New York: John Wiley and Sons, Inc., 1946.

and E. M. Kitagawa, eds. *Local Community Fact Book for Chicago, 1950.* Chicago: Chicago Community Inventory, University of Chicago, 1954.

ed. *Urbanization in Asia and the Far East.* Calcutta: UNESCO, 1957.

ed. *Population and World Politics.* New York: Free Press, 1958.

and O. D. Duncan, eds. *"The Study of Population": An Inventory and Appraisal.* Chicago: University of Chicago Press, 1959.

ed. *Urbanization in Latin America.* Ghent, Belgium: UNESCO, 1959.

Population Perspectives. New Brunswick, N. J.: Rutgers University Press, 1960.

and Beverly Duncan. *Housing a Metropolis—Chicago.* New York: Free Press, 1961.

and Leo F. Schnore, eds. *The Study of Urbanization.* New York: John Wiley and Sons, Inc., 1965.

et al. *The Future of the Family.* New York: Family Service Association of America, 1969.

ed. *The Population Dilemma.* 2nd ed. Englewood Cliffs, N. J.: Prentice-Hall, Inc., 1969.

and Evelyn M. Kitagawa. *"Differential Mortality in the United States": A Study in Socio-Economic Epidemiology.* Cambridge, Mass.: Harvard University Press, 1973.

5. THE FUTURE OF THE FAMILY

SUZANNE KELLER
Professor of Sociology
Princeton University

The subject for this evening is the changing American Family. The family is probably one of our few remaining idols and was for most of the people in this room the chief crucible of their character. We were all brought up in some sort of family, or wished that we were, and probably all of us have established some form of family in our own adult lives. So it must be quite disturbing to see the more superficial and sensationalist headlines about the disappearance and the decline of an institution that we had thought pretty permanent and that we care about a great deal. We *should* be disturbed. I don't think that casual prophesies and glib statements are what we need. My main hope this evening is to provide a more coherent framework for the discussion of basic trends and prospects for the modern American family.

As you will see, I think we're in for some deep and irreversible changes in the family in the next decades. I also believe that only by trying to understand what is going on, and by not hiding our heads in the sand, are we going to be able to cope with these changes at all.

And change, which so many fear, need not mean decline or decay, but may well lead to some needed improvements in family relations. Who in this room would deny that such improvements are in order?

Before turning to more substantive issues, I would like to state my definition of the family so as to avoid futile arguments over labels and semantics. The family means many things to many people, but in its most general sense it refers to socially patterned ideals and arrangements concerning biological and cultural survival. This has certainly been the historic role of the family. This is certainly what its role will continue to be, unless either biological reproduction or cultural transmission changes radically. Speculations to that effect will be touched on later. Let us agree here that the family is an institution with two fundamental tasks to perform: cultural and biological reproduction of the species.

I will start by trying to dispose of three common fallacies about the family. The first I call the fallacy of universality: It states that every society has a family similar, if not identical, to our own. All this statement accomplishes is to project our own patterns onto other people, thereby missing the reality of their lives.

The second fallacy is the fallacy of the family's biological inevitability, as if the family were caused by the biological features on which it builds.

And the third fallacy, the fallacy of antiquity, is that the family has always been with us and can be traced back to some ancient source. When we talk about "always," we usually mean two thousand years ago at best, a mere sliver of history. There is no unbroken continuity of the family from the Stone Age to the present.

Historic and anthropological evidence leaves no doubt in my mind that our family is a highly variable, "unnatural" (in the sense of being biologically determined) and very recent institution. In fact, *the* family doesn't exist. There are as many family forms as there are societies, and no single form is superior or rationally preferable to any others.

The only requirement is that the family be suited to the two purposes mentioned before, and that it do so with available resources. You build with what you have. And that's highly variable. Somewhere there is an adaptation, an interplay between the institution and the environment, but not a constant one.

The modern industrial family in its nuclear form, the one that interests us this afternoon, is actually less than five hundred years old, and historically rather unpopular. In response to many forces it is undergoing many changes, most of these irreversible. Far from being either natural or inevitable, it is in fact currently fighting for its life.

As for its antiquity, that too is illusory. If you take the famous Roman household to which we always compare ourselves, the patriarchy of the ancient Romans, you realize at once that in most important respects there is no similarity between the Roman household of several hundred people, including servants, friends, workers, apprentices, and close and extended relatives, and the three to four to five people we call the family today. It is a very different set of institutions. The Roman family was much more authoritarian, much more complex, much larger and not based exclusively on blood ties. In fact, the word family comes from the Latin term *familias,* which means household, and it really was like a small village, or a small community. To call these two institutions by the same term shows a certain indifference to detail and to their distinctiveness, which I think ill-advised.

Why do I harp on the variability of the family, its untypicality, its recency? Why do I care about that? Well, mainly because the fallacies have made it very difficult for us to recognize the legitimate diversity of family arrangements throughout the world. The fallacies are what Walter Lippmann long ago called "the picture in our heads," which we project onto the reality out there. But at the price of what? Of not recognizing that reality, of not learning from reality. And as a result we've been terribly impoverished about imagining alternative family arrangements and thinking up new ones. We've been blind to alternative possibilities, and this is unfortunate because there are changes ahead, and the only way we are not going to suffer too much from these changes is to try to anticipate them, give them a form, an image that people can respond to. I'm not willing to turn to false analogies of the family's antiquity, universality and inevitability to give us a false sense of security. In other words, I'm now going to look at our own time and our own family form with the proviso that it is highly limited, rather unpopular, rather recent, and subject to change, as all mortal things are.

So much for the preliminaries. In what follows, I intend to do two things: Review and summarize for you what I consider the basic trends and challenges confronting the nuclear family in our time, and then explore the implications for life, love and society for the next few decades. Let me begin by dividing the challenges to the family into two kinds: internal challenges and external challenges. They play a somewhat different role, and I think it is useful to treat them separately.

By internal challenges I mean all those strains and stresses that people experience as a result of family relationships, stresses that have little to do with the good will of the people and very much to do with the setup, the structure that's been created for them.

Now if we take an ideal definition of the family in the **United States** today, I think you will agree that it must contain the following things. The ideal family consists of a legally constituted husband-wife team and their young dependent children, all living together in a household.

This team is maintained by the husband's earnings as chief breadwinner and by the wife's concentration on the domestic scene and exclusive commitment to it materially and emotionally.

The first point to note about this "nuclear" family form is that it is very hard to achieve in reality, even when there are no great additional challenges from the changing times. It's very hard to achieve and always has been very hard to achieve, which may be one reason for its being rejected by the majority of the societies around the world. For one thing it's too fragile; more vulnerable to death and destruction than other forms, and also less capable of coping with them — because in its essence the nuclear family rests on the pair relationship, the dyadic, two-element relationship. And pair relations are inherently unstable — inherently, some would say, tragic. Because whether in chess, tennis or marriage, two are needed to play the game but only one can destroy it—by refusing to play, by playing badly, or by not being available. The departed or deceased husband, the missing parent, the incapacitated breadwinner or homemaker — each of these very well-known types in our own society are destroyers of domestic tranquility.

And then there are the internal contradictions of an institution in the throes of change. For one thing, there are obsolescent canons of authority. I'm sure most of you have confronted some of those in your own lives: authority relations between husbands and wives, authority relations between parents and children, which collide with a new ethic of togetherness, a new egalitarianism and changing ways of coming to agreement, of settling quarrels and the like. I think we will witness some profound readjustments here; the current generational confrontation is only a beginning.

In many ways the family as constituted now is the last remaining feudal institution in an industrial society. For in contrast to the achievement ethic of that society, the family is still based on ascription.

There is, secondly, the paradoxical role played by what I like to call the romantic love imperative. The romantic love imperative serves both as the prime motive for marriage and as the prime reason for divorce. And divorce is one of those contributors to the fragility of the nuclear family form which I mentioned above.

In other words, the nuclear family simply cannot weather the storms of emotional and economic crises in the way the extended family could and did. So whatever the virtues of this type of family, it also has some mighty problems, mostly due to its small size and its large responsibilities. For this reason, as you well know, a lot of its erstwhile responsibilities were taken over by other agencies. Think of education, think of health, think of old age security, think of recreation; all of these were once the province of closed extended kinship networks, and they are so no longer. But the trouble is that the agencies which were created as substitutes have stepped in half-heartedly and haphazardly because of our continuing prejudice that it really should be the family which does all these things, and our feeling that maybe someday the family will get back to doing them again. So we're caught in limbo between our belief system and the reality that this belief system cannot serve.

Thus, point number one, I'm proposing that because of the internal strains and difficulties just described, there is a latent reservoir for change in conventional family life in the most conventional sector of the public. That is, I am not talking about the people who can't fit in — I'm not talking about hippies, yippies and way-out people. I'm talking about very conventional people who are willing to support broader social changes without pioneering them. And it seems to me that we can't understand recent policy changes in abortion, day care, homosexuality, the pill, any of these things without realizing that the vast conventional sector was not opposed to them; none of these things could have been passed without their tacit approval.

So much for the internal changes. Let me turn to some external changes now. There are two broad fronts of change that I see combining with this latent reservoir of dissatisfaction, or of incapacity or frustration. One is a demographic change of great importance, and the other is an occupational change of great importance.

It is quite clear that in our time the reproductive span of a woman's life has been condensed enormously. If you compare, and I don't have time to do it now, but if you compare a twenty-year-old girl today with her grandmother at the same age and stage in life, you see profound differences as regards number of children, the extent of the life-span devoted to having children, the spacing of children and so on. Indeed, the whole absorption of a woman's life around the child-rearing function was virtually synonymous with her life-span.

Today the earlier age of marriage, fewer children per couple, closer spacing of children and the children's tendency to leave home earlier results in younger mothers and younger ex-mothers. Clearly, even if women play their roles straight, even enthusiastically, they are in for an unexpected and unpleasant surprise. If they're going to spend only some ten, fifteen years of an extended life-span on motherhood, what will they do with the extra decades? What do women do now with the extra decades? By their early forties they've done their duty to society, they've done the thing they've been primed for since, well, virtually since they were born, and then they have three decades of life left to be lived. What will they do with three decades if they're not prepared for them? So that's the first challenge. The "empty nest," as the demographers call it, is growing longer and emptier. And one thing you know if you study society is that things don't stay that way for long. Something will be pressing for change.

Now, if you go even further and take the idea of zero population growth seriously, and it has been taken very seriously by some, then an eventual reproductive ban is not far off. This means that women will have to find their prime meaning in life from things other than child-rearing and the domestic scene. And you see I am purposely focusing again on those women who would be perfectly content to play quite traditional, conventional roles.

The second main change that I want to touch on has to do with occupational trends. Increasingly in a society not prepared for it, women are being gainfully employed outside the home. But our current family setup is not geared to that; it is geared to their being inside the home the majority of their days. Despite this, some 40 percent of the women in this country have full-time jobs (in some age groups the proportion is much higher), and fully two-thirds of these women are married, and one-third have pre-school children. And I would like to know from you: In a society that prides itself so much on taking care of its children and the family as the key unit, how can you tolerate the fact that there are six million pre-school children of working mothers but only half a million places in day care centers? What happens to the five and a half million children not served by day care centers? Who takes care of them? The fathers? Older brothers and sisters? The neighbors? Other children? Or perhaps no one?

But why do women go out and work? Who are they? It's not just the minority of college-educated women, who see their training in ancient history being depleted by domesticity all their young, productive lives; that is a hardship, I know, but we're not talking about them. We're talking about women who go out to work because they must go out to work. A lot of them don't have male breadwinners;

they are either divorced or widowed. We can guarantee widowhood for most married women, since they marry men older than themselves and they marry into a category that is shorter-lived, so by definition they are in for widowhood. Another large part of the working women are married and are trying hard to provide the extras that are now considered necessary for a certain standard of living: health, recreation, and other things that the husband's earnings cannot provide. So the reasons for job holding are varied and not whimsical. And yet the majority of American women gainfully employed outside of the home are not recognized as job holders and must do double duty as domestic-wife-mother and as wage earner. Torn between home and job, between early childhood training and later reality, generally untrained and unskilled, hence forced to take work and wages not on their terms, these double-duty wives must manage as best they can in a society indifferent to their reality. It is *their* problem, we think. I don't think it is their problem. And very soon a lot of them will say it is not their problem.

This is why the women's liberation movement has gotten a lot more support than anticipated. It has come precisely from women who confronted the prejudices and traditions of this society and, I would say, its unfair treatment, and who suddenly become aware of a world that was not only not geared to them but outright hostile to them.

Now, lest it be thought that I'm one-sided in stressing only the effects of the trends on women, let me briefly take note of men's predicaments. They have their own problems — different, but no less real. Increasingly, men are beginning to question the reason for their having to assume, as a matter of course, responsibility for three to five people in addition to themselves. Increasingly, usually in complaints about the rat race, they're beginning to question why they should be working so hard, trying so hard, not relaxing, dying at younger ages than their wives, paying alimony as a matter of course, and deprived of their children in case of divorce. Now, these misgivings were always present among the non-successful males of this and many other societies. But they're now appearing as complaints or questions among the most successful, the so-called backbones of their communities. Confused by the generation gap, by women's protests, men have problems of their own. Authority and control—no matter how illusory control of human beings always is—were, after all, their rewards for their greater responsibilities. And as authority diminishes, so will the motivation to claim it.

All these trends would affect marriage, male-female relations, parent-child relations and work relations, even if there were no other developments on the horizon. But there are. Recent breakthroughs in biology, with their promise of a greatly extended life-span, of novel modes of reproduction and of dramatic possibilities for genetic intervention, will affect some of the most basic underpinnings of our traditional family arrangements. Let me just touch on a few developments. Genetic control, and sex determination of offspring, for one. Not just the prediction of offspring; we're on the threshold of determining the sex of the offspring. Already we can predict it—there are devices for doing so. Genetic surgery has become a recent new field. Genetic counseling. And so on. All these changes boil down to three things: One, human control over reproduction, which many used to think was beyond our control. Two, procreation without sexual relations. Three, sexual relations without procreation.

Now, once we've separated sex from procreation, each may develop independently. They need not be joined in one institutional pattern, as they are today in the family.

I hope I made clear what I mean by procreation without sexual relations. It means creating human beings by methods other than sexual intercourse. Artificial insemination is familiar to us already, but even more dramatic is the possibility of separating conception from gestation. And of course, all these things come into the culture legitimately, not by the back but by the front door. For instance, there are women who can conceive a child but cannot carry it. There are other women who cannot conceive a child but can carry it to term. If the conceiving woman has her fertilized egg placed in the womb of the carrier, you will have revolutionized motherhood. Where there was one biological mother role, here are two biological mother roles, not to speak of new legal and social roles. Sexual relations without procreation, on the other hand, means that sexual intercourse need not now lead to pregnancy and to conception. The pill and fertility control, of course, are the intervening variables there.

Having stated the case as I see it regarding three basic trends (demographic, occupational and biological), let me now try to spell out their implications. I have chosen to discuss three areas briefly: the area of marriage and family structure, the area of male and female roles and the area of sexual relations. I may not get to all of them but I will get to at least two.

What are the implications for the whole question of marriage and family relationships? Consider: If reproduction will at best constitute a smaller portion—let us say one-tenth or one-sixth—of one's life-span, and if sexual relations will not necessarily be linked to marriage, the question arises, Why marry, and why marry for life? Except for a small minority, this question could not be raised seriously before now. There were always a lot of people who did not get married, but they wished they were; they wished to have children. The ideal situation was considered to be marriage and family.

Furthermore, if a reproductive ban will replace the reproductive imperative by which we have lived for many centuries, then children will surely diminish as the prime reason for living. And for most people they are a prime reason for living. This is true for men as well as women. I know that men stress jobs and careers more, but they stress jobs and careers for the sake of what? For family, meaning children and their life chances. For women, of course, children have been the key commitment all along.

The reproductive ban must reduce the importance of the family and the salience of two roles now linked to it: the female-wife-mother role and the male-husband-breadwinner-father role. Surely we must redefine our notions of father, mother and child when reproduction is severed from sexual relations, marriage and conception. Who is the father? The progenitor or the provider, when these have become distinct roles?

Who is the mother? The conceiver or the carrier to term?

Who in fact will raise society's children under these conditions, and who will love whom? And by love, I simply mean assuming responsibility—toward whom? Love is a highly organized emotion today, though we think it's spontaneous. What new forms of love will develop to reflect new forms of life and of giving life?

The implications for male-female relations are great. If women are not to spend their lives in domestic roles, then obviously we will have to raise little girls very differently from the way we're raising them today: away from dependency and passivity, away from waiting for Prince Charming to pick up the meal ticket, and towards autonomy, self-reliance, activism and independence. I'm not at all sure that we're yet either aware of this or prepared to do this.

Recently I've had occasion to look into some very interesting material being collected on children's readers, and on how boys and girls are pictured in these readers at a very early age. It is really striking how early stereotyping starts, and how fixed those stereotypes are.

In addition, when women will have become self-supporting and in charge of their own sexuality, then surely our image of women and their image of themselves will change drastically. And perhaps as the traditional sex-work dichotomy break down or changes, we will regroup ourselves not according to gender, as we do now, but along interest lines, irrespective of gender. We're still not doing so as much as is possible, and as much as some societies do, but I've seen quite a lot of changes in that regard in the last two years. And as implied before, of course the male provider role cannot remain unaffected. The fact that child care, child support and wife support will take up smaller portions of a long life-span means very definite changes in the male role too. I don't see how the male can continue working for traditional objectives when these have been transformed.

I'm not even going to talk today about single parenthood and all these other possibilities. Perhaps they will come out in the question period.

Let's turn to some new experiments in marriage that have been occurring all around us; some of you may know of them, some of you may even have participated in them. I'm assuming you don't know many details about them, because nobody does yet, so let me run through some alternatives to conventional marriage forms.

First of all, there is plural marriage. I had a long discussion today with a group of twenty-year-olds, some of whom were trying to start a group family. That was all they wanted; they didn't want the nuclear nest at all. Now what plural marriage, with or without cohabitation, is designed to do is to bring back some form of the extended family, based not on kinship or blood relations but on interests. What this entails is a number of couples sharing life and life problems. There must be several hundred new communal arrangements consisting of adults, children, spouses, mates and singles, all trying to work out new patterns of life; some succeeding and, naturally, many failing. Even conventional family forms can fail drastically, so why should we expect that the new ones, without any traditions to go on, should prove themselves on the early tries?

The second proposal that I think you should consider, even though you might not find it appealing, is part-time marriages. These permit both togetherness and separateness at the same time. You probably all know of such arrangements, but only among certain kinds of people. Part-time marriage means simultaneously being together and separate;

the couple does not share one household but has separate apartments and meets at will. Now this would surely minimize the irritations of daily intimacy, which some consider an unfortunate price to pay for its blessings.

Then, of course, there are two-stage marriages; one has recently come back into the news via Margaret Mead, but it is really a very old idea. (In the twenties Judge Ben Lindsay of New York called it trial marriage.) Her proposal is something like this. Why not have two kinds of marriage: an individual and a parental form? The individual one is meant for the stage and age in life when you are experimenting, discovering whom you like and what you like. But once it gets more serious, she said, once you have a child or intend to have a child, then you have to opt for the second type of marriage, a parental marriage, which will force you to make commitments to that child and to that unit for as long as it takes to raise a child.

I have also seen an alternative proposal: that you have the parental marriage when you're young, and later, when you know yourself well and know your tastes and your style, you can have the individual form.

Then, of course, there is the possibility of no formal marriage, but of living together legitimately. If something is illegitimate, you don't feel right about it. Living together with public approval is what a lot of young people favor if they are not going to have children.

And then, of course, there is room for conventional marriage — but only, I would add, for the gifted few.

All these innovations seek to maximize two things that cannot be maximized today: security and variety simultaneously. Ultimately, of course, we may see more drastic changes. I think we will get to the point where, perhaps, only 5 percent of our society will reproduce for us. It sounds crass when put this way, but you know, once upon a time every family grew their own food. And now how many people grow food for the rest of the society? The answer is 5 percent. And I think eventually it may be same with procreation. Once upon a time everyone who could, reproduced for the total society. But now that we cannot afford this any more, it could very well be a specialized activity on the part of those who are gifted at it, who can do it well and who want to do it.

Now, there is no doubt in my mind that the transition through which we're passing will create many problems and leave many of us in limbo unless we prepare ourselves for it. Since most of us are still programmed for the old world, we are clearly candidates for culture or future shock, but already we see some positive steps being taken to work out new rules for ourselves, to write new scenarios.

At the same time, I think we must also realize that many of these novel immodest proposals which seem shocking when put as bluntly as I put them, because in a way, I want to provoke you, are already here, albeit they're not being focused on. Take part-time marriage, for instance. It shocks people when they first hear about it. But is this idea really so novel? Aren't such marriages characteristic of couples where the husband is earning his living as an airline pilot, ship's captain, traveling salesman, or any one of the numerous occupations that mean long separations, separate domiciles and periodic, if temporary, reunions between spouses and children? Such absentee husbands may be at home no more than three days out of every month. They don't raise their children on a day-to-day basis and they miss many family holidays. In other words, they're really not performing a very strong familial role or doing very much fathering. Still, they're a significant presence in the lives of their families, perhaps the more appreciated for their elusiveness.

Or take the sharing by couples of both domestic and non-domestic tasks. This is creating some hardship in American society, but you know, we're not alone. It's a big world we live in, and there are other societies that have managed to do this quite well. In the Soviet Union, Israel and Sweden, it has become routine to have husbands and wives share both aspects of life: jobs and home. Take Sweden, where it has recently been recommended that work schedules be so arranged that both husbands and wives can come home during the day and spend time with their children. There is, after all, no sound reason for holding rigidly to a nine-to-five schedule. Where does the nine-to-five schedule come from? From the agricultural era, I suppose, from a society of an era without electricity. But if you have electricity, why do you need a nine-to-five plan? You can divide and subdivide the day differently. By the way, in Sweden it is the domestic woman who is condemned, and the non-domestically-oriented woman who has her culture's support. A high proportion of middle-class wives there don't merely have jobs but careers, many of these in traditional male occupations.

At the same time, while motherhood is not neglected, it is not sanctified. Without the cult of motherhood, and with day care facilities, the conflict of career versus home which upsets so many middle-class American women has no place there at all.

It is not surprising, despite all sorts of economic inducements to have children (six months' paid maternity leave and other benefits for mother and children, whether they are legitimate or illegitimate—the distinction has been abolished there), that the Swedish birthrate is one of the lowest in the world. Swedes marry later than Americans, divorce less frequently and observe a single sex standard for the young of both sexes. The single standard removes a prime motive for marriage compared with the United States, where our double standard seems to be a principal reason for early marriages.

And as regards childlessness as a future mode of life, it is already here among what Richard Meier has called the sterile professions, such as truck drivers, diplomats and professional athletes ever on the move. Or among what I would call the narcissistic professions, such as creative artists of all kinds, in which one's progeny are, of course, the children of one's brain. It was Plato who made the distinction between the children of one's body and the children of one's brain. And Francis Bacon once wrote that the care of posterity lies in those who have no posterity. What he meant, of course, was that the problem-solving activities of mankind, the thinking, creative activities, have been carried out by people who did not have to worry about sustaining day-to-day life and taking care of numerous progeny.

Clearly, some significant portion of the population can lead happy but non-reproductive lives in a biological sense. Those among us who fear for the moral if not the biological survival of our species, if these innovations come to pass, need not be alarmed. Sexual permissiveness, now condoned in many countries or parts of countries, need not lead to license. A recent comprehensive study of sexual attitudes in Sweden by Hans Zetterberg should lay all fears to rest. Sexual permissiveness has obviously increased in Sweden, but there is no evidence of sexual profligacy or sexual promiscuity. These are themselves, of course, terms with a Puritan flavor. On the contrary, devotedness, concern for the partner, mutuality and generosity of spirit mark the responses of the latest and sexually freest generation. Fidelity and trust, disapproval of adultery and betrayal, were strongly, one can say overwhelmingly, endorsed. Indeed, one feels that some of the innovations we've discussed may promote a respect and sustained interest between married partners that is all too rare in many long-term marriages today. Some uncertainty may actually be a very good thing. As in genuine friendships, it would be better if you couldn't take each other so readily for granted. In this light, consider a recent proposal for yearly renewal of marriage licenses. I don't know how that would sit with you, but it is designed to keep alive a sense of challenge and interest in a relationship which a feeling of effortless permanence apparently erodes all too easily. After all, says one proponent, we do as much for drivers, voters and dog owners.

Similarly, I think, it would be useful for us to have a good comparative look at the successes and failures of collective institutions for child care. We're so used to American individualism, the individual family, the individual nest, the individual childhood nest, that I think we're missing out on some of the advantages of these. I'm not sanctifying or idealizing them. I've looked at them myself, I see the problems. I don't think any society, industrial society, has solved any of these problems satisfactorily. But there is something to be said for growing up with your peers at your own pace, for growing up with voluntary mothers, for growing up with a much greater sense of community responsibility and community concern. In many ways the children fare very well. I realize it's a controversial subject and I certainly wouldn't want to treat it lightly. Above all I am not suggesting that inadequate communal facilities are desirable, just as I don't think inadequate parental facilities are desirable. But there are adequate communal arrangements which we should consider.

So let me, in conclusion, sum up what I consider fairly safe speculations about future developments in the family sphere.

First of all, a trend away from family concerns as a major lifelong commitment for men and for women. This trend is a result of the reproductive ban and increased longevity, as well as the emergence of new cultural priorities. I think the frontiers are going to lie elsewhere; not just in space, but in new kinds of problem solving. Please note that I said "lifelong commitments." Family life may thus be one coveted stage of everyone's life.

Second, I see a trend towards greater legitimate variety in marital and sexual relations.

Third, a decrease in the negative emotions now all too often associated with romantic love and parental love—exclusiveness, possessiveness, fear, jealousy and anxiety about love relations.

Fourth, greater room for personal choice in the kind, extent and duration of intimate relationships. This enlarged choice may greatly influence the quality of the relationships, since people will both give more to them and expect more from them.

Fifth, more experimentation with extended non-family forms of living. New kinds of groupings of people based on shared interests.

Sixth, multi-stage marriages and quasi-marriages geared to the changing life cycle and the presence or absence of dependent children.

And seventh, some new principles of human association not geared solely to family and kinship concerns.

The question arises whether most people that you or I know — whether, indeed, most people in this room — would respond positively to these innovations and changes. Observation and evidence from a few surveys suggest to me that people may be more ready for change than official pronouncements, and above all many "experts," assume.

The spread of contraceptive information and the acceptance of fertility control have been nothing short of remarkable. Erstwhile taboos have disappeared within our lifetime and public forums no longer shy away from such vital but previously forbidden topics as abortion, homosexuality or illegitimacy.

Moreover, the better-educated, who are on the increase, are more receptive to change in all these respects.

A Louis Harris poll, as reported in *Life* magazine, showed that the majority of Americans did not reject even such startling intrusions as egg transplants, test-tube babies and cloning. They did not seem terribly surprised by these innovations, but they were hoping to use them in the service of a better family life.

For the immediate future, most Americans, young as well as old, opt for and anticipate their participation in durable, intimate heterosexual partnerships. While these relationships would continue to be anchors and pivots of their adult lives, people also expect them to be freer, more flexible than they were in the past, and less bound to duty and involuntary personal restrictions. This is for the immediate future. For the long-range future, I think far deeper changes may be expected, but that's something for another session. So I will stop now and it's your turn after the coffee break. Thank you.

SUZANNE KELLER

Suzanne Keller, currently a professor of sociology at Princeton University, is known for her work on elites, community planning, the family and futurism.

Born in Vienna, Professor Keller came to the United States as a child and received her secondary, college and graduate education in New York City. After college she spent several years in Europe, mainly in Paris and Munich, where she worked as a survey analyst and translator. Upon her return she attended Columbia University and received a Ph.D. in sociology in 1953. A post-doctoral fellowship took her to Princeton's Center for International Studies for a year, and in 1954 she became a research associate at MIT, specializing in survey design and analysis. In 1957 she accepted an appointment at Brandeis University as assistant professor and there taught courses in social theory, stratification and the sociology of religion.

In 1960 Professor Keller returned to New York to work on a book and in 1961 joined the staff of New York Medical College to concentrate on problems of social class and intelligence measurement. A year's teaching at Vassar was followed by a Fulbright lectureship in Greece in 1963. Her assignment, at the Athens Center of Ekistics directed by C. A. Doxiadis, marks the beginning of her interest in architecture and community planning. At the completion of her Fulbright in 1965, Dr. Keller joined the Doxiadis staff, where she worked on various projects, gave lectures, and supervised research until 1967. That year she came to Princeton University as a visiting professor, and in 1968 she was the first woman to be appointed to a tenured professorship there.

Today, Professor Keller is pursuing her varied interests in teaching, writing, research, public lectures and world-wide travel. A federal grant is currently permitting her, and several other members of an interdisciplinary team, to investigate methods for the assessment of planned environments. She is also active in the Women's Rights Movement. The author of numerous articles and two books, she is currently completing a book on changing sex roles in America.

BIBLIOGRAPHY

"*Beyond the Ruling Class*": *Strategic Elites in Modern Society*, rev. ed. New York: Random House, 1964.

"*The Urban Neighborhood*": *A Sociological Perspective*. New York: Random House, 1968.

6. THE SURVIVAL OF MANKIND

WILLIAM W. BALLARD
Professor of Embryology
Dartmouth College

I am not here to predict the future: you've had that already. You have discovered that there are optimists and pessimists, prophets of boom or doom. The highest-paid professional soothsayers seem to be the ones who will give you your choice of forty different futures, merely listing the innumerable variables in the unwritable equation. I content myself with reviewing the course, to date, of Dr. Alan Gregg's planetary disease, "world cancer." Several years before he died in 1957 this very distinguished medical educator contributed a paper to an AAAS symposium on population problems in which he developed a metaphor in some clinical detail: "The world has cancer, and the cancer cell is man."

I propose to develop his forensic diagnosis as a biologist sees it, defining the symptoms and course of the disease and commenting on what has to be done if it is to be arrested. I have the unquenchable optimist's belief that the disease, though far advanced and potentially fatal, is curable. To hear it proposed as national policy that we now start forking out whatever funds are asked in the anticipation of a quick cure of human flesh cancer sounds to me brashly political. The world cancer, on the other hand, is not so mysterious. We know what causes it, and techniques for its cure are at hand, however distasteful they may be.

First I have to deal with some formal concepts. Invite a professor in, that's what you get. Let us start with some exponential curves (see Charts A and B). As you know, you can get different-looking graphs for the same equation by changing the units on the X and Y coordinates. Changing the compound interest rate up or down steepens or flattens the curve. Graphs like these warm the hearts of politicians and economists when they show the growth of per capita income, demand for electric power, productivity in an industry, history of the GNP, etc., etc. Biologists meet many such curves also, but see them as temporary episodes in cyclic or recurring events, signs of imminent danger or change, impossible to project in the long run. Sooner or later, depending on the rate of the compound interest, these curves approach infinity at unimaginable speed, which is impossible in the real world.

EXPONENTIAL CURVES

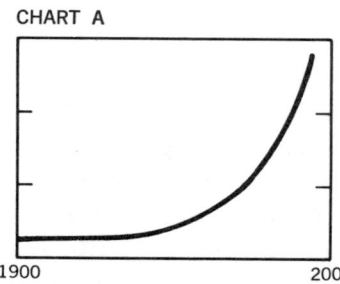
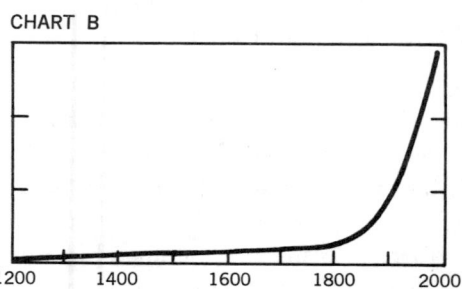

As an example, my beer crock. A college professor who likes his beer compromises with ultimate realities by brewing his own. I put a little yeast in a 5-gallon crock of well water with some sugar and malt extract, and overnight it bubbles. Next day there's hissing and turbulence and a meringue floating on top. It's simple to estimate the doubling time of the yeast cells. After something like the number of doublings that the human population has made since written history began, the rate changes in a way predictably correlated with overpopulation, exhaustion of natural resources, and pollution—if the word may be applied to alcohol. Unless I bottle the batch there is presently a yeast population crash, and a total shift in the biological nature of the contents of the crock, leading to a swift decline of its value to me. For an exhilarating time the yeast cells were multiplying on an exponential curve, and they "never had it so good." If they had been a voting democracy, warned of the leveling-off and the imminence of the crash (a and b of Chart C), they would have voted to continue the joy ride, throttle wide open.

That's what a biologist learns to expect from a closed system. With its limited and unrenewed resources, my beer crock is a sort of closed system. In a vastly more complicated sense, so is the earth. More like a natural biological system is the apparatus called a chemostat. The beer crock becomes a chemostat if you add a constant controllable inflow of nutrients and a balanced outflow of yeast crop and pollutants. Regulating the factors, you can maintain a constant yeast population per unit volume at any desired level on the curve, such as K_1, K_2, K_3, taking off from the growth curve, which would otherwise be sweeping up logarithmically (Chart D). Once the controlling factors have been manipulated to the desired point, the chemostat has, within insignificantly variable limits, a fixed *carrying capacity* for yeast.

An actual self-regulating biological system is very much more complex, so stupendously complex that concepts, language and tools for studying it have only recently been invented. A lake, a pond, a river, a ditch, a field, any small reasonably uniform living community on the face of the earth—what we call an *ecosystem*—consists of very many

BEER CROCK

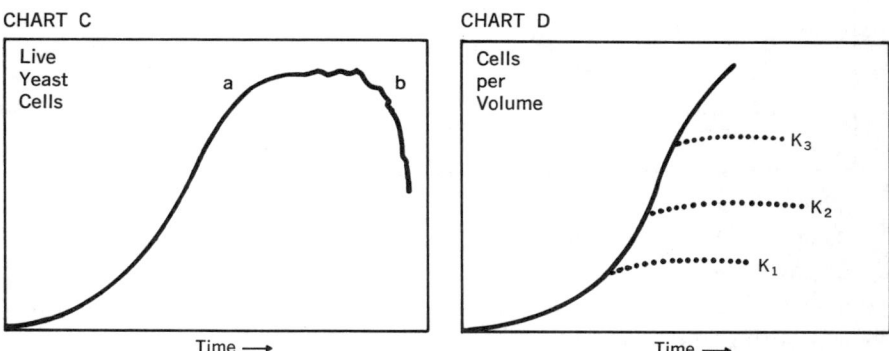

thousands of species of organisms, each a population in itself, all engaged in the exchange of materials and energy among themselves and with the non-living environment.

There are three main classes of organisms in any such ecosystem. First the *plants*, which receive the energy of sunlight and use some of it to build themselves up from water, carbon dioxide and a lot of other essential elements and compounds. Second, the *animals* that build themselves up by eating plants or each other. And lastly a very large, very numerous class of *decomposers* such as bacteria and fungi, which break down the substance of plants and animals, returning their elements to the environment. Each of these three classes provides opportunities, and sets limits, for the other two. The recycling of essential substances makes possible a very much larger standing crop of living organisms at any given time than would be possible if they all had to be created out of virgin materials and freshly arrived sunlight (Chart E).

The ecosystem not only continuously provides opportunities for each of its multitudinous species, but sets limits upon them. Each environment provides for a relatively fixed number of members of each of

ECOSYSTEM

CHART E

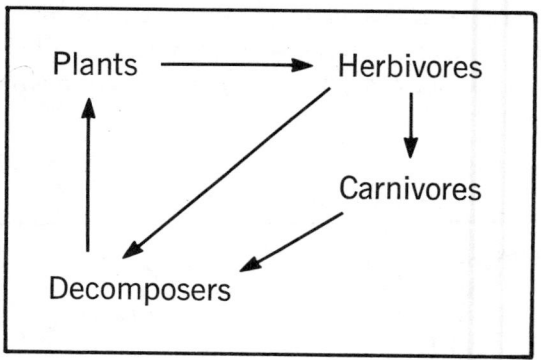

its species. In stable condition, it has a definite carrying capacity for its carnivores, its herbivores, its bacteria, its trees, its snails, its wildflowers, balanced out over the seasons and over the years. This was one of the crucial insights of Darwin himself.

Curiously, Darwin was tipped off to this by a minister. It was the Reverend Thomas R. Malthus who pointed out that the human population grew geometrically while the food supply increased arithmetically. Food then became a limiting factor, i.e., fixed the carrying capacity of agricultural acres for many. It was the Malthusian conclusion, the famous "dismal theorem" of economics, that if people would not themselves control their numbers, their numbers would be cut off at some fixed level by starvation, wars, pestilence and other miseries. Darwin saw similar limitations pressed upon all species, all organisms, with the pressure hardest on the less fit individuals.

Though this led directly to the history-making Darwinian theory of the mechanism of evolution, the point here is that the concept of carrying capacity has been accepted in biology for much more than a hundred years. Food is not always the limiting factor that sets the capacity for a given species in a given environment. It might be water, or oxygen, or temperature, or sunlight, etc., etc. But a limit normally exists. As soon as a species exceeds its limit, feedback mechanisms appear that cut it back.

All the carrying capacities for the myriad species within a given environment are interlinked by food chains, energy exchanges and other ways, and this may be demonstrated for a particular environment in a general way by the existence of a "pyramid of numbers": the number (or tonnage) of individuals turns out to be proportional to areas of compartments in such a diagram as illustrated in Chart F.

Bringing these introductory theoretics to a close, let me reassure you that my intention is to look at the progress of Gregg's world cancer. I can now postulate that the disease started when prehistoric man first discovered how the pyramid of numbers could be manipulated in his environment so as to increase the carrying capacity for himself. Malthus and Darwin were right; every living organism is programmed

PYRAMID OF NUMBERS

CHART F

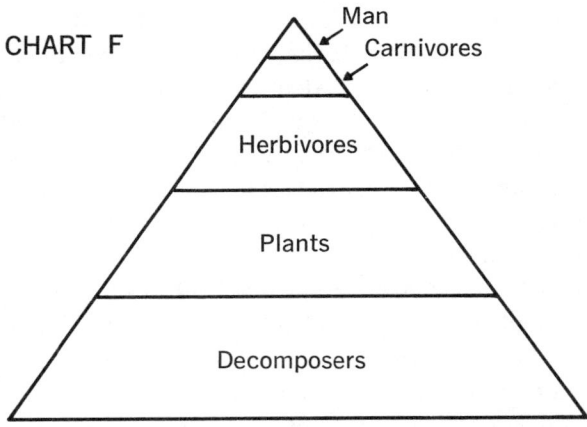

so as to reproduce its own population up to the point where some limiting factor forces a halt. Human beings, and very few other organisms indeed, have discovered ways of releasing themselves from such a bind. In fact, human history is a Perils of Pauline serial of escape from one restraint after another. Each new technological advance, each release, has seemed a blessing and has been immediately copied far and wide, and each time the population has built up to the cut-off level of some other limiting factor. For ten thousand years we have been in an accelerating race between technology and the Malthusian trap.

As a consequence, man has been the only glaring exception to the rule that the environment will support a relatively fixed number of individuals per species. The national census is a recent invention, but an educated guess is that ten thousand years ago there were five million living people. By 1650 there were probably five hundred million. The population had doubled five or six times, at an average rate of once in 1500 years. The next doubling, to one billion people, took 200 years. From 1850 to 1930, or 80 years, we doubled again. We may be four billion people as soon as 1975—doubling time, 45 years. The rate of doubling for the world population currently averages once in 35 years or so.

That's logarithmic growth for you. It has become a game now to invent demonstrations that it can't continue. Go 800 years forward, for instance, and you have to accommodate the people with standing room only, five square feet each, *land and sea.* If the present growth rate had started and continued from the time of Christ, there would be twenty million times as many people on the earth as there now are. Such growth can be supported only for a few generations before the feedback controls take over. The biologist sees the growth of the human population as not only unprecedented but unsupportable.

The trick has been pulled off by three equally revolutionary types of human behavior which I would like to examine in turn: (1) swift adaptation to new environments by social rather than biological means, (2) huge-scale elimination of competing species, and (3) the use of extraneous sources of energy.

Each of these types of human behavior has involved the invention and exploitation of technologies. The economist Kenneth Boulding has introduced us to an "utterly dismal theorem" that follows if the "dismal theorem" of Malthus is right: If there is an inexorable growth of the human population toward the level of starvation and misery, then the ultimate effect of any technological advance is to increase the total number of people who will eventually get caught in the Malthusian trap.

SWIFT ADAPTATION TO NEW ENVIRONMENTS

If the archeologists read the still fragmentary evidence right, the first human being and the first technologies of fire and weapons and food storage evolved in tropical Africa. The early populations huddled in very small groups in mighty safe places, and must surely have been held in rigid check by carnivores, starvation and disease. From Africa they spread to Europe and Asia, and then by way of a long-lost land bridge to Alaska and the Americas. By the beginning of written history, before the great age of exploration, people were on all the continents and all the big islands, making a hard living by special techniques of tool-making, shelter-making, hunting and food-raising. Every new environment they moved into presented special dangers and special resources, and they had not only the daring and the drive to move in but also the intelligence to adapt. This could not have happened quickly, and it produced some astonishingly different cultures. It is hardly possible that in North America before the arrival of the modern Europeans, a tribe of Eskimo could have changed place with a tribe of plains hunters, or a tribe from the desert mesas of the southwest could have changed place with a fishing tribe on either coast. None of these transplanted tribes could have survived a year, so highly specialized were their technologies.

This restlessness and adaptability to country and climate did not greatly affect the Malthusian limiting factors themselves, but it released the species from the limitations of the original habitat on the African plains. It undoubtedly permitted many doublings of the worldwide human population.

ELIMINATION OF COMPETING SPECIES

Without ever having heard about the pyramid of numbers in his ecosystem, prehistoric man began to discover how to manipulate it to his advantage. Consider New England over the last four hundred years. One published estimate is that the area was occupied by only about 5000 Indians. They were principally carnivores, though they did make small clearings in the forest and raise a bit of grain. They were preyed upon, and competed with, other carnivores such as wolves and wild cats, including the great catamount or panther. The forested plains and hills were almost useless to them except as producers of deer and bear, birds and other small game. Every winter these Indians were in danger of starvation. Food was probably the main Malthusian limiting factor.

Since then we New Englanders have eliminated most of the other land carnivores, and we have preempted for ourselves the space at the top of the pyramid of numbers that once represented their population, their tonnage of living material. We have manipulated the herbivore part of the pyramid to our advantage by limiting the numbers of certain undesirable species (woodchucks, muskrats, rabbits, insect pests) and introducing more desirable ones (sheep, cattle). The increase in biomass of herbivores serving as food for man directly increased the biomass of people supportable by the ecosystem. We have further manipulated the plant part of the pyramid to our advantage, by cutting down forests and draining fields to make more cropland from which more man-food in the form of edible grain and meat-on-the-hoof could be raised. Each accomplishment along these lines was a controlled alteration of the carrying crop for particular desired species or pest or weed, which in turn boosted the carrying crop of people.

This can be done without enlarging the entire crop of living material in the ecosystem. That is, the triangle representing the pyramid of numbers remains the same size, though the compartments inside are being readjusted to make a bigger area for man. But slowly over the centuries ways have also been found to increase the productivity of the land, i.e., to enlarge the triangle itself. Great agricultural inventions like plowing, fallowing, irrigation, fertilization and selection of better genetic strains of seed and livestock have actually changed the levels at which some limiting factors begin to operate. For instance, in my vegetable garden one of the principal limiting factors is water supply in July and August. Supplying the water increases the crop, which then becomes limited by the usable supply of nitrogen and phosphorus. Supplying these will further increase the crop, which is then limited by something else: temperature? daylight? disease?

Every local human culture was finding its own effective ways to manipulate the ecosystem, and each such advance eased off the threat of starvation for a time. The living was easier and safer. A slightly higher percent of babies lived on to the breeding age and the population crept up to the next Malthusian limit. Exploitation of the environment became more efficient as people specialized: homemakers, farmers, toolmakers, warriors, rulers. Communities grew and cities began to appear, though the death rates must have been very high in them. The limiting factors of predation and starvation were probably replaced by disease. No doubt the early cities grew more by immigration than by reproduction. The medical history of Europe down to modern times is a ghastly repetition of famines and epidemics, all the evidence one should require for Boulding's "utterly dismal theorem."

EXTRANEOUS SOURCES OF ENERGY

Biological systems seem to be an exception to the thermodynamic principle that matter tends to disperse in chaos, losing its structural complexity and organization. Ecosystems tend to greater complexity, but only because there is a continual input of energy from the sun. Their constituent organisms maintain and reproduce themselves by using, exchanging, and wasting this energy. The early advances in agriculture resulted from the application of human muscular energy to production of food from otherwise useless land: lugging water, hitching yourself or your woman to a plow, and weeding. Later, of course, people harnessed draft animals and made them do it. These were in fact enormous advances. Boulding suggests that the invention of the horse collar may rank in importance with the invention of the wheel. Locally wind power and water power came under control to release manpower for more subtle uses, and there were always productive or protective benefits that eased off the desperate race between human reproduction and human death and allowed the population to grow on, toward a somewhat higher Malthusian limit.

But then in the eighteenth century the invention of the steam engine, in the nineteenth century the invention of the internal combustion engine, and the rapid exploitation of fossil fuels for the energy to run both of them, blew the lid off. As late as 1850 it was reasonable for a New Hampshire farmer to reply, when told that the population was going to double, "It's impossible. Where'll they find the pasture for all their horses?" Since then non-animal sources of power have been applied to every phase of agriculture and to all the industries and other occupations producing materials and things that contribute to the production of food. Optimists began to say—and there are still optimists with this curious blindness to the general biology of populations—that food need never again be in short supply for man. The Malthusian theorem was dismissed as inapplicable to us.

It is the conclusion of ecologists who are now studying the flow of energy and materials through ecosystems that the enormous new productivity of agriculture is due more than anything else to the input of fossil-fuel energy into the system. We are, approximately, exchanging calories of fossil fuel for calories of food. In this country and in much of the world, the enormous recent increment in human population is being fed, by indirect means, from coal and oil and natural gas. The level at which food would now become a Malthusian limiting factor has been raised so high in this country that the danger is incalculably small, and the shocking fact of starvation in our midst is at once recognized as a maldistribution scandal. Meanwhile, the doubling time for our population continues to decrease. This is the effect not only of our productive efficiency and richness in natural resources but also of health technology which rescues our young from the childhood diseases and brings most of our adults far past maturity.

I contrast here two kinds of curves that could represent the history of the human population. The first (Chart G) is rigged to show how recent and how fast the increase has been. The second (Chart H) is my own interpretation of how man has escaped successively from one limiting level to another by his technological advances. Each time, he rigs his own chemostat to a new carrying capacity (K_1, K_2, K_3 in Chart D). Then, maybe before or maybe soon after he reaches the hard times where his population comes into balance with available resources, he finds out how to shove the limiting level up again. I have named a good many of these techniques (increase of cropland, use of fertilizers, domestication of herbivores, application of extraneous power, etc.) but the chart does not list them or give historical dates because its function is merely to present the biologist's general interpretation of what is happening.

Let me come back to the main thread, following the course of this

HISTORY OF HUMAN POPULATION

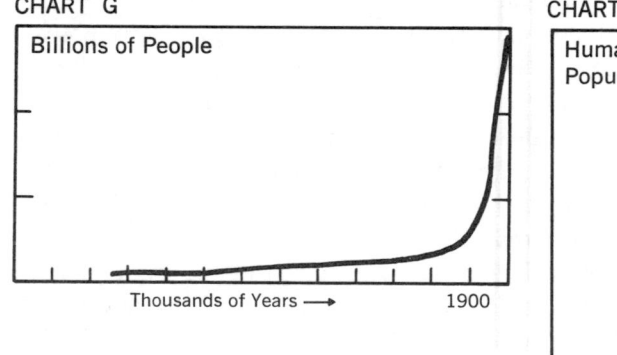
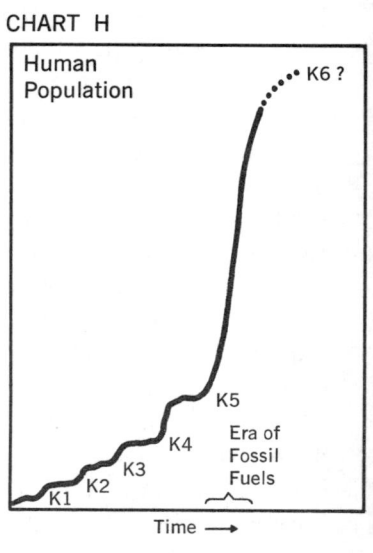

planetary cancer. The disease-producing organism, man, in exploiting new freedoms discovered in his behavioral virtuosity, has set aside one biological restraint after another. His multiplication has gone wild, almost completely out of control. As in flesh cancers, the causative units have also broken away from their locus of origin, wandering over the earth, invading almost all the kinds of land environment, preempting and disrupting and destroying the naturally evolved structure and organization of local ecosystems. Spots of man-made desert remain to record where their self-seeking zeal went too far and broke down the living tissue of the biosphere, like the necrotic spots of flesh cancer.

I'm going to venture the thought that there are already on the face of the earth about five times as many people as can be optimally supported on a long-term basis. The point is certainly arguable, perhaps we will come back to discuss it. However, my objective is to review the course of the disease.

In any ecosystem, including those preempted by man, waste heat (the eventual exit of energy from the system) cannot for long exceed the energy being taken in, or the system will run down. There is similarly a budget of materials, such that the system cannot use more physical stuff over the long haul than can become available, or pile up more by-product and waste than can be dispersed and recycled by suitable mechanisms. These are external restraints on the ecosystem, above and beyond the requirement that its internal economy for recycling energy and materials shall be in balance. If one species, such as man, ties up an increasing amount of energy and material in its own expanding population, there must be viable internal compensations in the ecosystem, or else new sources must be found. Conversely, if man turns out more waste heat and material for dissipation than natural mechanisms can cope with, another set of problems arises. These in fact are the symptoms of the world cancer that is being caused by the proliferation of man and his industrial society. I review now as briefly as possible the shortages coming up, and the mounting excesses, both of which measure the extent of the disease.

SHORTAGES COMING UP

Food

Documents regularly brought up to date by the United Nations and various national government agencies remind us that close to two billion of the nearly three billion people now alive are in a state of chronic malnourishment. This indicates that while the world population is doubling in the next 30 or 35 years, we should be multiplying our food production by three. Optimists hope that the predicted century of mammoth famines can be held off long enough to achieve ZPG, zero population growth. Pessimists, admitting that we have the technology to do this, find the slight signs of current progress both in food production and in birth control to be inadequate by whole orders of magnitude, and remind us of the staggering political, economic, social, cultural and even religious barriers that have to be overleaped in concerted international effort if the goals are to be reached.

I am willing to accept expert opinion that we already have the technology to double the world's food supply in short order. Cheap journalism keeps giving us puffs about hydroponics, farming the oceans, making cropland out of the Amazon basin, etc., but these are delusions of ignorance. One source of encouragement is the large crops of wheat, rice and corn now being harvested on certain tropical lands through the use of new seed strains and heavy investments in machinery, fertilizer and irrigation. This "green revolution" requires and may not get adequate support from industry, finance, transportation and firm government policy. It may fall prey to devastating blights adapting to the new strains, or to social unrest or war. In any case, the most serious need is not for huge grain harvests but for animal protein, the lack of which is at the root of much of the worldwide malnourishment. Protein deficiencies, easily recognized in acute stages, stunt the body, the intellect and the spirit, particularly during childhood, before they are detectable.

In just the countries which are reaping the big new crops of grain, the population is expected to double in thirty years or so. The mind boggles at tripling agricultural produce before then, and keeping abreast of such a torrent of demand thereafter. And if the outcome of our best agricultural technology will ultimately be "utterly dismal" because of uncontrolled population growth, what choices should we be making now?

Energy

Depending on a person's size and activity, the energy needed to keep him healthy and growing is about the same as that needed to keep a 100-watt bulb burning. Up until quite modern times man took in this much energy in the form of food, and perhaps used another

400/500 watts of energy derived from firewood and other sources. In this country at present we use something like 10,000 thermal watts per capita, which is like having nearly a hundred personal slaves apiece. Some of this energy is converted to space heat and lighting, much more into the power of steam and internal combusion engines, and machines run by electricity. The demand for electricity has doubled each ten years for the last thirty. It rose 9 percent in the last single year for which I have seen figures, 1968-69.

Part of this extraneous non-animal energy produces the luxuries and comforts to which we have become accustomed, but a huge slice of it is required for raising food. One has to count not only the mechanical power applied directly to the land, but also the energy used in building and maintaining the dams and pipes to collect and deliver the water, transporting the supplies such as fertilizer and pesticides, operating the factories and mines that produce these and other equipment, and still further, processing and distributing the food itself. We have added to the human ecosystem a unique mechanism for increasing the carrying capacity for man: we buy a vastly increased food supply by the expenditure of fossil fuels. We could not revert to the agriculture of the hoe and the horse without a drastic cut in the population, or at the very least the uprooting and dispersal of most of the city dwellers by the threat of starvation.

Yet the fossil fuels, on which we chiefly depend for energy in this generation, are non-renewable resources. The National Academy of Sciences, through its Committee on Resources and Man, warns that production of natural gas in the United States (barring Alaska) will decline starting in 1978 or 1980. In the years 1950 to 2015 we will have used up 80 percent of the non-renewable supply. It is more difficult to estimate world oil reserves, but a similar curve of expanding production, to the year 2000, and subsequent rapid decline, seems to fit, with 80 percent of the once-only resource burned up between 1955 to 2025. There is a lot of coal left, but within our own lifetime we will be down to much more expensive and much nastier grades of it.

Hydroelectric power accounts for less than 2 percent of what we are now using, and in the United States practically all the best places for reservoirs are already occupied. Siltation will diminish their usefulness to zero in a century or two. Geothermal, tidal and solar sources of extra power are being explored with new interest, but there is no present hope that they can substitute for more than a few percent of our reliance on fossil fuels.

The hope of the future, of course, is atomic energy. The uranium used as fuel in the few commercial atomic power plants now operating will not work unless it is enriched slightly with the naturally radioactive isotope Uranium-235, which is so rare that it may be exhausted for practical purposes within a couple of decades. Therefore these plants will have to be replaced quite soon by the liquid metal cooled fast-breeder plants, which will make more fuel than they consume. But these are still only in experimental design and are sure to be vastly more dangerous. It is simply not known yet whether atomic fusion can be controlled for commercial power production, though there are optimists as well as pessimists among those wrestling with the incredibly complex physical problems. If and when solutions to these problems are found, the stage of commercial exploitation would presumably take another three decades.

Projecting the curve of demand for electricity into the future on its present equation—and you recall that all this can do is to convince you such growth cannot be tolerated in the real world—it is estimated that in less than two centuries the entire continent of North America would have to be covered with thousand-megawatt generators, leaving no space for transmission lines or people. A more imminent slide-rule horror is that every bit of running water in North America will be needed to cool our electric generators by the year 2000. In the meantime, amongst the local power failures and brownouts, the demand for power continues to soar, and in whatever community it is proposed to install a big new generator, there is at once a public outcry of dismay and wrath because of the atmospheric and thermal pollution and visual ugliness that will come with it. Particularly, the legal techniques for forestalling or eternally delaying the installation of new atomic power plants are now elegantly developed. This country is not in danger of the great famines predicted for the impoverished and overcrowded parts of the world, but the problem of sharing out the power in this power-mad industrial society of ours may be the biggest Apple of Discord ever.

Other Non-Renewables

One can't have an industrial society without *metals*. Of the 100 most needed metals, this country is now importing some proportion of about 90. We import more than 50 percent of the following, according to 1967 figures: antimony, asbestos, bauxite (for aluminum), beryllium, bismuth, cadmium, chromite, cobalt, industrial diamonds, manganese, mica, nickel, platinum, silver, tantalum and tin. We go farther and farther for these things, using ores of diminishing quality. The new high-temperature technologies, such as capture of energy from atomic fission and, we hope, atomic fusion, will make us excessively dependent on some quite rare and relatively unheard-of minerals. One cannot generalize that as these metals become more rare and expensive, substitutes will be found. There are important exceptions to the generalization that "given cheap and unlimited power" we can go on mining less and less rich ores. Some metals do not have that kind of statistical distribution.

Critics of economic imperialism point out quite rightly that we here are monopolizing vastly more than our share of these limited world resources, and thereby diminishing the possibility that the so-called undeveloped countries can have an industrialized future. It has been calculated also that if the rest of the world were by some miracle brought up to our level of affluence by the year 2000, the annual production of copper, iron ore, and lead, for instance, would have to be multiplied by 11, 12 and 16 respectively. If this were possible for one year, it certainly would not be possible for the long haul. Even if the workable ores and the power held out, the lid would be held on by the resulting thermal and chemical pollution.

There is plenty of *water* on the earth, but practically all of it is salty or frozen. The fresh and liquid portion amounts to only 0.04 percent of the total, and most of this is underground. To live within our income, we should limit our use of fresh water to the amount that falls on the land as rain. We already use a third that much, drawn from our rivers and lakes, and demand is going up in the usual alarming exponential curve. What happens when the population doubles and redoubles in size? In addition, we are mining the ground water at a much more rapid rate than it is replenished.

The course of ground water toward the sea is very slow, so that in a given locality it should be considered a non-renewable resource. Over-pumping it from wells has dropped water tables out of economic reach in a number of heavily populated or desert areas in this country. And over-pumping is drawing salt water into previously fresh water wells in some coastal regions such as the California Bay Area and Long Island. Estimates of water requirements in the United States in 1980 considerably outreach the optimistic estimates of what will be available. Interstate and international squabbles about the rights to water in particular rivers become more and more shrill.

In pursuit of more usable fresh water, Andean glaciers are being melted back by sprinkling coal dust on them, but the practice can be predicted to have disastrous side effects in the end. The mining of polar ice is not an economically attractive solution for most parts of the world. Given cheap and unlimited power—that much-dreamed-of variety of pie in the sky—one could supply coastal cities and irrigation systems through desalinization of sea water, but if this were practiced on a large scale the obvious by-product would be an embarrassment. What do you do with a cubic mile of salt? Fresh water may become a limiting factor worldwide, as it now is locally.

Soil on which crops can be grown is formed very slowly indeed by natural processes, so that it should also be considered as a non-renewable resource. The United States has led the world in soil conservation practice, but we are still losing out by erosion and over-exploitation. We succeed in maintaining the per-acre yields from much of our richest land by application of more and more fertilizer to compensate for the hidden damage. By malpractice we continue to produce more desert than is being reclaimed for us by great irrigation projects. In the long run, irrigation itself often turns out to be a self-defeating practice because of slowly developing degradation of soil structure and composition.

With all our technology we in the United States still have much to learn about how to practice a really permanent agriculture. In other parts of the world, particularly the overpopulated and tropical countries, examples of marching deserts and abused or ruined cropland are much more shocking. But because the hungry people are already there, it is neither economically nor humanly possible to take the pressure off the soil and encourage restorative processes.

THE EXCESSES

In this category I lump the problems which arise, not because the earth cannot give us all we want, but because the earth is being messed up with the by-products of our activity. The present uproar about pollution is a sign of growing awareness of the problems. The development of technology to heal over these cancer spots is well in advance of its application. The hard part is getting general public willingness to enforce corrective measures and pay the cost of them.

You people here must be well informed on *air pollution*, connoisseurs if not aficionados. It is power that pollutes. The smoke and soot, the poisons, sulfur dioxide, carbon monoxide, the oxides of nitrogen and their photochemical by-products, nearly all come from our engines, as they furnish those vast amounts of power by which we manipulate our ecosystem to support our bulging population. Engines can be made more efficient as energy converters, but the more efficient they are, the more they pollute. To substitute non-polluting electric motors for internal combustion engines merely concentrates the source of pollution in the mammoth generating plants, as well as increasing the demand for scarce metals. We can ask Con Ed and the other unavoidably conspicuous culprits to burn less poisonous fuels, but this merely forces our more numerous descendants to use the worse fuels later. There is technology for scrubbing the soot and poisons out of the smoke at the stacks, but this usually substitutes pollution of water or land for the air pollution. Power pollutes, and we continually demand more power. That dome of smog that covers every city in the world and drifts out across land and sea is a part of the cost of our way of life, and an obvious symptom of the world cancer.

Water pollution is of many kinds, each requiring its own technological remedy. Human sewage, agricultural runoff, and food-processing wastes create related problems. We have a long way to go in installing the old-fashioned primary and secondary treatment plants, but even these are inadequate solutions for an expanding population. That crystal-clear, odorless, apparently drinkable water from a good secondary treatment plant may still be loaded with enough nitrogen, phosphates, pesticides and other pollutants to damage a river or ruin a lake.

Industrial effluents were about as poisonous or revolting—except in quantity—a hundred years ago as they are today, but for a long time now they have swamped the recuperative powers of nature. Industry has been alert enough to its selfish interest to make sure that government agents are thwarted even in their attempts to estimate the size of the problem. As the clamor for control rises, so does the blackmail cry, "You'll force us out of business! Unemployment!" Meanwhile the deterioration of our lakes and the even more disastrous condition of our estuaries is being documented. Blame cannot be fixed on any one factory or corporation, and practically nothing is known about the technology or the time required for recuperation.

Some industrialists, after listening to enough flak about what they were visibly dumping into the rivers, are discovering that the cheapest way out is to drill a well a thousand feet deep and slyly pump their poisons underground. This alarming trend has not had enough public discussion. No one knows where the stuff goes, or how fast, but it cannot do otherwise than contaminate the ground water, a precious and practically non-renewable resource. In the few horrid cases where such poisons have come to the surface or turned up in people's well water in a spreading area, it has been difficult to fix blame, and impossible to work out a remedy.

Thermal pollution of lakes, rivers and estuaries by using water to cool power generators and other mammoth heat-wasting industries has caught the ecologists with inadequate knowledge. Slowly accumulating after-effects are difficult to guess, and prior conditions are seldom documented in sufficient detail, however clear it has become that when man-made manipulation of the temperature of a body of water is begun, wide-ranging effects on its ecology will slowly appear. It is like performing surgery with the eyes closed. This is an issue that is hotly debated locally when each new mammoth power station is proposed.

Solid wastes now present a very difficult problem for every city and town in the industrialized world. In the United States we are rich enough to produce far more than we need, import what we lack, and waste lavishly—even unwillingly through planned obsolescence—but nobody will take our trash and garbage in return. Burning it on dumps pollutes the air. Burying it wastes unreplaceable land. Protest is finally mounting against dumping it in the ocean. Technology is being rapidly developed and very slowly applied along more sensible lines. Garbage can be burned to generate electricity at competitive cost. Paper, glass, metals and many other substances can be salvaged and recycled. Each of these processes produces its own undesirable side effects, of course, and often must be promoted at public expense.

Among the most extensive and intractable cancer spots are the mountains of mine tailings and square miles of debris from strip mining, and the areas in and around former mining towns where coal is burning underground or the land is sinking as old tunnels collapse. Each generation has presented the next with huge and unavoidable costs that have to be met by the general public, since private enterprise did not include them in the price of the product it was selling.

Radioactive wastes already present a frightening problem which will very rapidly get worse, and for which no good long-range solution seems in sight. The leakage of radioactive gases from power plants is a measurable though not yet a worrisome hazard, but it is a complex and dangerous business to dispose of the principal radioactive fission products of the atomic piles. In this country we have large deserts and abandoned salt mines to bury the stuff in, but other countries are not so fortunate. Great Britain is still dumping in the ocean, as we also have done in the past. We have completely reliable information about the half-lives of the innumerable by-products that must be kept away from people and out of the ground water, but we know little about the half-lives of the containers we are using. If you need more to worry about, there are the dangers of highway accidents in transporting such material to the processing plant or to the graveyard, the possibilities of disruptive earthquakes, or direct bomb hits in a war, etc. We are told we need not worry about the danger of one of the atomic power plants getting out of hand, but one hears less of the possibilities of "cheap limitless power" as more is learned about the cost of built-in safety controls. There seems general agreement that it will be more troublesome to control our thirst for more power than to go on multiplying the problems that are created by power production.

My list of the things we have too much of ends where it should have begun — with *people* or "peopollution." It is people who generate the shortages of food, power, metals, water, and the lot. They are the sources of the pollutions, of air, water, heat, radioactivity, and the mountains of trash, the symptoms of the world cancer. Without straining the welcome of your guest biologist I cannot go on to comment on the unmeasurable but real dangers of simplifying the human ecosystem through our control or extinction of other species. This is most clearly seen in the increasing vulnerability of our agriculture to new pests and diseases, and the discovery of hitherto unimportant limiting factors in our manipulation of the living world. I can only mention the long-range danger of the now detectable biological concentrations of pesticides, toxic metals and other poisons up through the aquatic and terrestrial food chains that support us. I am incompetent to discuss with you the things we are doing to the atmosphere and to the oceans. These are now beginning to be monitored through international scientific cooperation because of the danger that we may be initiating irreversible trends affecting the ability of the planet to support human life, or life itself. Let us return to more immediate concerns.

It is often suggested that the proliferation of people, and their apparently uncontrollable concentration in cities and supercities, is by itself and irrespective of their polluting by-products, a part of the world disease. Are the crowding, the noise, the economic cannibalism of the city adding to the self-destructive and earth-destroying potential of man? There are some famous and suggestive experiments on crowding of rats, with profound effects on their social organization and individual behavior. Whether crowding of people has similar effects, and if so what result this might have on human evolution, can only be subjects for speculation.

Nevertheless, signs of disorganization and pathological function in human society do exist and may be increasing. Recent events here in New York have shown how vulnerable you are to the disruptive capacity of a few small groups of people pursuing personal gain without regard to community welfare. A handful of people can reduce the city to chaos in a few hours in any one of fifty ways. The revolutionists study the vulnerability of your bridges, aqueducts, power plants, subways and communication centers, and the corruptibility or distractability of your public servants, with a fascination we should all share. You live in as much danger as the Athenians in the time of Thucydides. You will get faster and more massive help in a real crisis than Athens got, but when the comparable emergency comes, vastly more people will die.

We do not lack for suggestions as to how to cure the internal rot of the cities. With all our open spaces, why not split New York up into forty or fifty small cities, and scatter them out there where there's nothing but corn and cows and cotton and stuff? Somehow providing water and sewerage and power and industry and transportation, and overcoming legal obstacles and human resistance. I leave this to you. The main source of the world cancer is too many people. Scattering the cancer cells merely sets up new centers of proliferation.

We already have quite good technology for controlling the expansion of the population and gradually and painlessly diminishing the pressure of people on their ecosystem. Biologists and chemists are rapidly improving that technology, but the big problem is consumer resistance. The Planned Parenthood people are making some progress with the easy part, which is to prevent the birth of unwanted babies, but the hard part remains after this objective is reached, because apparently people will still want too many babies. Even if the more ambitious goal of ZPG is reached in 30 years, the leveling off will leave us with at least 50 percent more population than we have now. And every added American will monopolize as much of the world's resources as 25 or 30 people in India. Obviously it is up to us to stabilize our own population before preaching to the rest of the world.

But one can still be a determined optimist. When things get sufficiently bad we will come to our senses, pull ourselves together, and find a way on. As Thornton Wilder had it in his play *The Skin of Our Teeth,* we have always managed to squeak through. I think we here will have a rough time learning to live with less gasoline and electric power and waste and water and babies, but we will not suffer directly the great famines that will cut back the overshoot of population elsewhere. We will get to recognize the signs of the environmental damage all around us, make a rational diagnosis of our disease, and work out a sensible course of treatment. The world cancer can be cured.

WILLIAM W. BALLARD

William W. Ballard has devoted much of his professional life to research in developmental embryology and comparative anatomy, work recognized by a series of grants he was awarded by the National Science Foundation. But he has also had an interest in the broader implications of science as part of a liberal education.

He graduated from Dartmouth in 1928 and accepted a teaching fellowship at Yale from 1928 to 1930 while he did graduate work. He received his doctorate from Yale in 1933. In 1930 he joined the Dartmouth faculty, and became professor of embryology at the Dartmouth Medical School in 1942. Professor Ballard has spent many summers in research at the Marine Biological Laboratories at Woods Hole, Massachusetts, and for several summers he taught there as an instructor in the embryology of marine animals. He has also done research at the Stazione Zoologica in Naples, Italy, and the Station Biologique in Roscoff, France. He has developed methods for a curricular approach to biological research at the undergraduate level and has written a textbook of vertebrate morphology.

In pursuing his interest in science as part of a liberal education, Professor Ballard served on the committees which set up the Great Issues Course at Dartmouth as well as a course in the history of science. He served as director of the Great Issues Course in 1948-49. He has written numerous articles for scientific journals and is a member of several learned societies, including the American Society of Zoologists, the Society for Developmental Biology, and the Society for the Study of Evolution.

He has also been active in civic affairs, serving the town of Norwich, Vermont, as president of the development association, a member of the finance committee, and a school director. He was one of the leaders for Norwich in the establishment of Dresden District, the first interstate school district in the country, and served as one of the first chairmen of its board of directors.

BIBLIOGRAPHY

Comparative Anatomy and Embryology. New York: The Ronald Press Company, 1964.

Articles by William W. Ballard usually appear in biological journals covering the embryology of lower vertebrates. Three on salmon embryology are soon to appear in the *Journal of Experimental Zoology.* Others of recent years:

1968 "History of the Hypoblast in *Salmo.*" *Journal of Experimental Zoology.* 168: 257-272.
1969 "Normal embryonic stages of *Gobius niger jozo.*" *Pubbl. Stazione Zoologica Napoli.* 37: 1-17.
1970 "Archenteric Origin of Midgut Lumen in Amphibia." *Developmental Biology.* 21: 424-439.
1970 "Horsemen of the Apocalypse." *Massachusetts Review.* Autumn. 781-809.

7. HOW TO THINK ABOUT THE FUTURE

JOHN KETTLE
Writer and Consultant

Mr. Chairman, ladies and gentlemen. Someone said — and I don't know who it was—"If you don't think about the future, you can't have one"; and that seems to me a good starting point for thinking about the future.

Everyone must be more or less a futurist at some level of his life. There's a story about a deep-sea diver who had scarcely reached the bottom when a message came from the surface which left him in a dilemma: "Come up quick," it said. "The ship is sinking." Now, here's the futurist: the past is sort of sliding away and tomorrow is coming in, and you don't know which way you're going. You'll see this a little more clearly, I hope, later on. You're thinking of the future as an extension of the past, and somehow having to relate it to what you're doing now. If futurists do nothing else, they give you some way of understanding where we are today.

You've probably noticed that as soon as a new highway is opened, it's jammed, it's clogged. In other words, the planning has gone wrong. Somewhere between the drawing of the first line and the opening of the road, the curve went a bit higher, more people bought more automobiles and decided to make trips in them, and the whole new highway system is too small. And I've heard that New York Telephone had similar problems, which simply indicates that you have to think about the future in a very practical way, or you can't even go on in the present.

So let me say something general about this. In the past when people looked to the future, they expected things to stay about as they were; and things did. Later, people expected things to evolve in smaller regular steps; let's say they expected the population to increase by twenty million a year, and to go on increasing by twenty million a year, in a straightforward way.

The great literature on which we were brought up gives a sense of stability, almost of stolidity, which is very deep in us and which has formed our notion about what the future is.

But now in the past thirty years something quite different has happened. We've learned that things don't go in straight lines. They increase in a compounding way; 5 percent a year or 10 percent a year. They are accelerating steadily; not growing steadily. This has made people scared of understanding the future.

So now we have a situation in which a futurist like Olaf Helmer, with the Institute for the Future in New England, can say: "The future is no longer viewed as unique, unforeseeable and inevitable. There are instead, it is realized, a multiple of possible futures, with associated probabilities that can be estimated and to some extent manipulated."

This is a new idea. I mean, it is an innovation in the whole business of looking at the future.

Now, if, as I was, you were brought up by parents and teachers who were born at the end of the last century or the beginning of this, and if you looked at what was happening at that time, when progress consisted of twenty million a year or 2 percent on gross national product, or 2 percent was the interest paid on bonds, you understand what they meant by progress. Those are the ideas that were put into you. Yet you know that your children are growing up in a world in which changes can be 7 percent a year or 10 percent a year, like the gross national product of Israel; or 15 percent a year, like the increase in air passenger miles; or 20 percent a year, which is just a little more than the annual growth of the gross national product of Japan. Now, these are fantastic changes. In a very few years something quite different happens. You get a change in quantity so great that it is in effect a qualitative change.

If you have a 2 percent change every year, the thing you're looking at doubles in about thirty-five years; that is to say, in about one generation.

If you have something that is changing at 8 percent a year, you have a doubling time of nine years. Now, what are we to say— that nine years is the new generation? Or if you have something that changes at 12 percent a year, doubling time is six years—is that the new generation?

Well, curiously, it's something like the university generation, and this may be a key.

Consider that a young man who enters the teaching profession today will influence children who will be alive and active as adults a hundred years from now. So it is important to get these things into some sort of perspective in ourselves. We all are to some extent teachers, and for many of us, part of our job is actually to teach professionally.

So I say that there is in the world we live in an atmosphere—or, if you like, a feeling—that continuous development on all fronts is the normal condition of our lives; that it is the regular—or you might almost say unchanging—background against which our personal, corporate and public decisions must be made. The unchanging background is change. That sounds like a paradox, but I think you understand what I mean.

Well, there it is, and one recognizes that one has to do something about preparing for it.

I was talking to a group of people upstate one day, and there was an insurance man in the crowd. We were talking about social change and how it affects the corporation, and he got up and said: "I used to know how to deal with change in the corporation. The other day I went into the john and there is a guy smoking pot. What do you do?" And there were tears in his eyes, he was so baffled. "Do you send him home for the day?" said this fellow. "Do you dock a day off his wages, or what do you do?" . . . Complete bafflement.

Well, everybody thought this was kind of funny, and I mentioned the story at another event and said: "Well, what do you expect?—insurance companies are very, very slow; they don't know how to deal with these things. We're all intelligent people. We know what to do if there's a guy smoking pot in the john." And after the meeting a man came up to me and said: "I'm with an insurance company and you're right; but we really are now snapping to attention. The whole thing has changed a lot. We are beginning to solve this problem. We've got a whole new image—the with-it, now-generation image."

"Yes?" I said. "Yes," he said, "We have a vice president in charge of interpersonal relations"—I think was the term. And I said that sounded nice. And he said: "What's more, we've redesigned the company flag."

I don't think that's the way to face change.

Let's say the problem of looking at the future is really two problems: How to be bold enough to imagine a world that has not yet happened. And how to be sensible enough to put your imaginings about this still-nonexistent world into a time and space context. So I say flatly, we are not going to be vacationing on the moon in the 1970's, and I say that for reasons I'll elaborate on a little later. I say we may well be using computers at home in the 1970's. There are attempts to draw a distinction between what may and may not and what can and cannot happen, and I think we have to do this sort of thing.

If we look at the future we have to admit that what we can know about it is almost nothing. We can know the maximum number of people age thirty or over in the year 2,000, because they've all been born—a fact—maximum number. We can know the movement of the stars, because this is a huge mechanism with an enormous momentum that it would be impossible to stop very rapidly. We can know about the growth and decay of living organisms and of minerals. We can know that human nature does not change very much—hardly at all in a lifetime. So that's what we can know.

But we can't know much else. Given that very scanty amount of knowledge to use in looking at something as weird and as complex as our future is going to be, how can we do it? How can we think about this future?

Well, let me mention a number of techniques. Some of them will be familiar to you: probability statistics—the Monte Carlo game. Some of you who are engineers, or depreciation people, or so on, will understand those. The delphi technique. When you came in here tonight there was a questionnaire; you checked off some answers. If we were going to play the delphi game right, we would put on the board the consensus of your answers. Then you would look at these answers and you would say: Well, I didn't agree with the consensus. And you would answer the question again, explaining why or changing your mind. And then this can go around. You can put your objections and the other people can read them. And so gradually you can either come to a consensus, or polarize. This is one way of trying to get a common feel for what might happen.

Mathematical projections or geometric or logarithmic projections of various sorts—computer simulations, computer models—some of these, again, will be familiar to you. Writing scenarios—this is something the Hudson Institute has done very well. You imagine a situation, let us say, twenty-five years in the future. That's your end point. You write, as it were, a drama which explains how you get from here to there. You look at it and you criticize it for inconsistencies, improbabilities, incredibilities, and other things.

Historical analogy is another very good technique. You can say: Well, there was another country that got into this sort of jam a hundred years ago, a thousand years ago, and what happened was, they did so-and-so. Is that what's going to happen to us?

Innovation in itself will change the future, and in itself will stamp its mark on the future—although I will qualify that later on.

And the other thing is to be a genius, and disrupt the future. Be an Einstein; understand and explain that $E = mc^2$, and change the future in a way which only you would have understood.

Chart 1 was made after a delphi study, and it has to an extent reached consensus. First the chart gives the record from 1930 to 1965, or thereabouts—total national expenditures for education as a percent of the gross national product in the United States. Up to the late fifties around 3 to 5 percent of the sum of goods and services produced was spent on education. By 1965 this figure was around 6½, 7 percent.

CHART 1

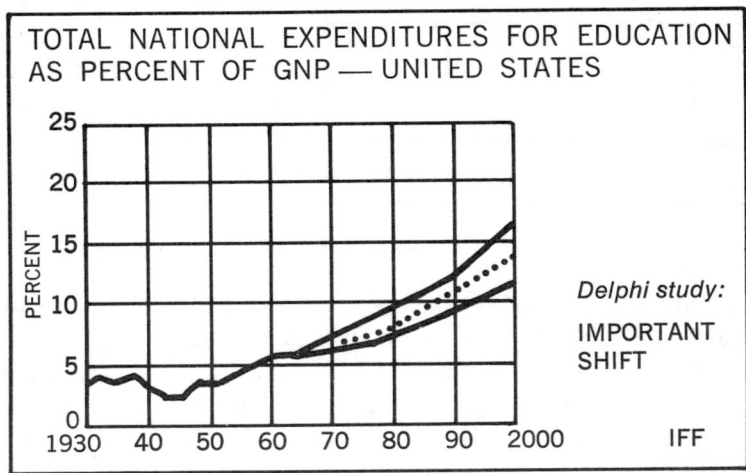

Now, then, in the delphi technique a lot of people are asked what's going to happen: Is it going to go up? Is it going to go down? How fast? When? And so on. And the two solid lines on the chart starting from the mid-sixties are the end result of three, four, five rounds of delphi questioning.

These people predicted an important shift, as the chart shows, from about 7 percent to about 15 percent; in other words, doubling in a generation, thirty-five years. There is a disagreement, but not a very wide disagreement, and the dotted line shows you what the average would be. Okay.

Chart 2 gives the results of another delphi study; this one dealing with U.S. government spending on foreign aid. It started in about 1945, which would be the Marshall Plan, with 6 billion dollars, which in those times was a lot of money. Then it sort of went up-down, up-down, up-down, until in the mid-sixties the spending was about 5 billion.

CHART 2

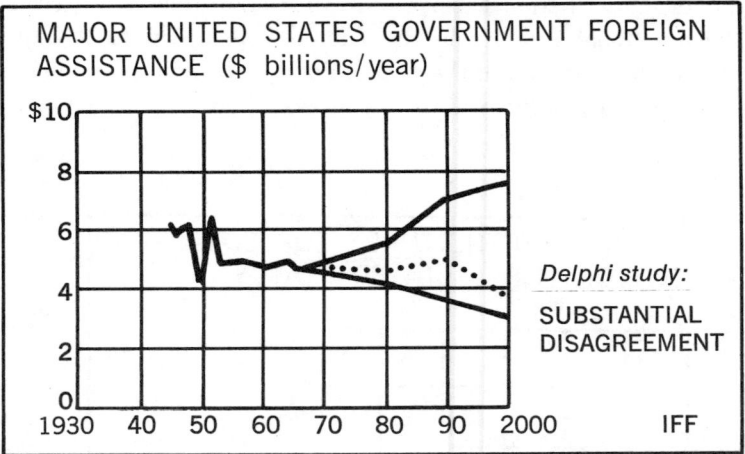

The experts and the not-so experts were asked: What's going to happen next? And here you get, as the chart says, substantial disagreement, a polarizing. Some people said it's going to go up, and they may have had very good reasons for thinking so. Some people said it's going to go down, and they may have had good reasons too. If we were playing the game ourselves, we would want to discuss the reasons. In this particular one, which was conducted by the Institute for the Future, some people said, "It's got to go up because the poor nations will put an enormous amount of pressure on us—psychological pressure, moral pressure or even military blackmail." You know, some of the underdeveloped nations can have the bomb. China is one that does. And they can say: "Give, or we'll blow you to hell." Other people said: "What will happen is that the United States will become increasingly isolationist, and no threats, no moral persuasion will have any effect. And foreign aid will go down to 3 billion and will keep on going down." And, of course, compared with what GNP is going to be by the end of the century in the United States, that will be a very small sum.

So these are the sort of results you can get out of a delphi study. Now, of course they don't tell you what's going to happen, but they give you some reason for starting to think; also, of course, a basis on which to propagandize, if that's what you intend to do.

So there are some techniques. Now I'm going to start giving my talk—because all that was a preliminary. I want to give you ten points, ten ways of looking at the future or ten things to do when you're looking at the future.

1. *The Mechanics of Change*

Grasp the mechanics of change. Well, I've talked about that to some extent. Let me talk about it a little more.

We're accustomed to saying—or some of us are accustomed to saying: Here we are in 1970. We're halfway between the year 1940 and the year 2000. Sometimes it's put in another way. Think of the changes that have occurred since 1940 and imagine all of that happening again, or something just as drastic happening by the end of the century.

Well in fact, that is underestimating change. In our world, change occurs faster than at first sight seems reasonable. If a phenomenon is changing at 5 percent a year, you haven't got a straight line; you have a curved line. Then we're halfway between 1940 and 1990. The amount of change that took place in the last thirty years will occur in the next twenty; that is, if change is regular at 5 percent a year. If you think about a phenomenon that is changing at 25 percent a year (and there aren't very many of them), we are halfway between 1940 and 1979; in the next nine years, as much change as in the last thirty. One such phenomenon is the increase in the number of computers; that rate of change is about 25 percent a year, and has been for nine years.

If you think of a phenomenon that is changing at 50 percent a year—and there are very few of those, but one is the increase in available computing power—we are halfway between 1940 and 1972. That is to say that all the change which took place in the lifetime of some people present in this room can take place in the next two years.

These figures are very hard to grasp. I mean, it sounds like magic. But you can go and do the numbers and see how the thing works. So I just say, understand that change can occur very fast. Even with a 5 percent change a year, we are halfway between 1940 and 1990.

On the other hand, many new technologies are at the beginning of their development and will peak out before any such thing happens. If we look at a child in the first five years of his life, his typical rate of growth is 22 percent a year. If he went on growing at that rate, by the time he was twenty-one he would be thirty feet high. How many people do you know who are thirty feet high? So there is a "technology"—human growth—which does not continue at a steady rate, but accelerates and decelerates. Many natural phenomena go like that, and we are beginning to think that many unnatural or man-made phenomena will go like that too. We'll start slowly. We'll spurt. We'll cool off. Some technologies will have to follow that pattern: we now know of at least three—maybe more— each of which at its present rate of increase would soon consume more than one-third of the GNP. That would be impossible. The rates at which we're going in for computers, welfare spending, education, radical transportation systems are examples. There are others too which could very easily consume a third of GNP before the end of the century. Well, you know that's not going to happen, because we still have to buy clothes and bread and roofs.

Let us take a look at something which you've already given some thought to. Will the Japanese be wealthier than the Americans, and when? Here are the facts. Japanese per capita income is now about $1700 a year. American per capita income is about $4700 a year. So Americans are 2½ times as wealthy as the Japanese. But Japanese per capita income is going up at about 15 percent a year. U.S. per capita income is going up at about 5 percent a year. Now, you can simply draw curves and see where they intersect. Since it's tiresome to do it in your head while someone's talking to you, I will give you the answer. It is 1983.

Now, of course it's perfectly reasonable to argue that before then the U.S. economy will hot up to, let's say, 7, 8 or 9 percent, and before then the Japanese economy will cool off to about 12, 11 or 10 percent. The intersection will then occur later, and we shall see how many people have done this rather more sophisticated thinking and have come to the conclusion that 1990 was the right year, But on the present information that is before us, we say 1983.

So those are the mechanics of change. That's the way they work, and it's not everybody's baby. But you can do it, and it isn't terribly hard. I mean, I'm a journalist and I can do it.

2. *Technology and the Real Force for Change*

Think about technology. We all have a tendency to think that technology changes people. But remember that technological innovation does not work unless people want to use the thing you've just invented and put on the market. And in fact, I will say dogmatically: the great force for change is a deeply and widely felt desire, though it may well be an inarticulate desire. You have had the experience, I know, of people not using technology for what you intended. Telephone booths are an example, and there are other things you can think of—technologies which have been invented and put on the market only to be ignored, or else to be used for something quite different. This demonstrates to us that technology will work if it does what people want it to do, and wanted it to do before you invented it, practically speaking. Let me give you a couple of examples of technologies to think about for a minute.

I believe that in the communications business, one which interests most of us, there is a future technology which is going to link the computer and print and television. Let me say something about what is happening in each of those sub-systems, in each of those bits . . . just throw a few bits out and see if you put them together in your mind and come up with a picture.

Anyone, a private citizen or a corporation, can buy into the *New York Times* Information Bank and gain access to all the information in the *Times* library. And as they build up their clientele in this thing, they will plug in other libraries, as they are put into the computer, and eventually into graphics and eventually into other sources of information.

The director of the U. S. Information Agency about a year ago was talking in public—and this is not a guy who's given to making wild, hairy statements—about a world information grid in which man's entire knowledge would be available to everyone on demand. Anybody in the world could find anything that was known anywhere in the world.

Think about the way your kid flips TV channels. I watch my boy, who's eight, go to the TV and he goes click, and he looks . . . "hmmm!"— click; "hmmm!"—click . . . "boring!"—click . . . "rerun!"—click . . . very fast. Kids can absorb a whole lot of information. Now, mind you, television is not a high-content medium at the moment, but it's interesting how fast relatively untutored children can spot what's going on, make a relatively complex decision that they would prefer to look at something else, retain some idea of whether they'll come back to Channel 9 or not and so on. It's interesting, because it shows something. It shows that they have this desire—well, they have a lot of desires all mixed up in there, which they will not articulate to you. But watch them develop. They want a variety of things. They want a breadth of information. They want to be able to switch among the channels. It's in itself very interesting for them. They want it fast. I still don't know the statistics for the United States, but in Canada we now have 300 cable TV systems in operation. This is compared with 100 daily newspapers.

How long will it be before I can dial direct to India? I can go into the *New York Times* Information Bank. I can get from some cable TV systems twenty-five different programs. I will soon be able to get on the Xerox system an instant print of a rare book. I will almost immediately be able to get a sound and image cassette to plug into a home TV.

Well, now, this suggests a sort of amalgam—and I've only mentioned a very few things; you know a whole lot more—which is building into something that looks very different from the type of communication system we have right now.

Some of you will have heard of Gordon Thompson at Northern Electric, the man who invented the scribblephone. Thompson does not like Picturephone for reasons which I would prefer him to tell you; he thinks that it's not a very great advance on what you've got right now. His scribblephone is a system whereby you can interact with the guy at the other end; you can play tick-tack-toe. You can write on your screen and it shows up on his; it changes what's on his. He can write on his; it changes what's on yours—and so on, backwards and forwards. Interaction. This is a very important development, I think, and I think Thompson has drawn out of the sort of items that I've been telling you about a sense of what people are looking for when they as kids flip the TV dial; or when as adults, they buy into the *New York Times* Information Bank.

So it seems to me that quite soon the computer promises to become a data picturephone, plus memory. And if that's so, then the future of the telephone is not very large unless the telephone happens to be the thing I'm talking about. Watch your dust; I'm ready to see you do it. But I don't see it yet.

Let me talk about another technology: town planning. The whole thing is wrong, this concept that you can plan a city and thus make people do what you want in the city. Cities grow because they attract people. That's what I have to call Kettle's platitude; you know, it's a very straightforward thought, except that no planner seems to have understood it. He seems to think that you can adjust the city in some sort of impersonal way. Well, if people express their desires, they'll just simply say the hell with it, and go live somewhere else.

We see things like Megalopolis building up—you know, the Boston-Washington complex down the east coast, which has forty million people living cheek by jowl. Why? Because people want to live this way. Okay, what are people looking for in the cities? They're looking for opportunity and they're looking for variety and they're looking for sociability. And you can get all that. But they're also, and this is interesting, looking for privacy and they're looking for order and they're looking for efficiency. And you can't get any of that, you know, particularly in New York.

So somehow or other you have to think the thing is going to collapse. You have to think the cities are about to degenerate. And then you look, and lo and behold there's Detroit over the horizon, sort of liquifying before your very eyes. The last time I was in Detroit, there was grass growing on the sidewalks down the main street. It's an incredible situation. In one of the best parts of town, a woman I know went out to bring in the laundry from the yard, and took a walkie-talkie with her with the key open. I do not understand why people will live in cities like that. The answer, I think, is that they won't, and so we have to imagine at least the possibility of a sort of optional city or a nucleated Megalopolis which will really be sprung apart. You cannot govern a city of forty million people from the mayor's office. You have to think of something with a lot of autonomous units which come together on an optional basis.

Let me say about technology that it won't always do what you expect it to do. You know that Henry Ford when he invented the car advertised it in this way: "The car will enable a man to enjoy with his family the blessing of hours of pleasure in God's great open spaces." What has done more to destroy the great open spaces than the automobile? There is one technology that simply didn't do what it was expected to do. And there are others.

3. *Experts—and How to Do It Yourself*

Listen to experts, but be skeptical. Professor Bickerton, who was an astronomer but who is remembered today largely because of this remark, was talking in 1926 about the "foolish idea of shooting at the moon . . . the proposition appears to be basically impossible." What was Goddard doing at that time?

In 1956 a fellow countryman of mine, the British Astronomer Royal, Dr. Woolley, said: "Space travel is utter bilge." What were they doing with Sputnik in 1956? The damn thing was practically ready to go. So be skeptical about experts. Be skeptical about second-hand reports. Try to go to real sources; go to the astronauts rather than to Professor Bickerton or Dr. Woolley. Be skeptical about statistics. Try to get statistics from the people who are actually doing the stuff.

I would say, do your own mathematics, if you can. Do not rely on other people, because other people generally get them wrong. And if you get them wrong, at least you've had the fun, and it's also quite easy to go back and see where you've got it wrong.

Let me combine this thought. The experts say we're in for a leisure problem in the future. In fact, a lot of people are making a very good living by going around telling us so. Yet we have not got a leisure problem, as far as I can see. We have, maybe, a work problem. We do not have a leisure problem.

My grandfather worked for sixty to seventy hours a week. He started off at, I think, sixty-six and he came down to sixty-two hours a week, when the employers got kind. He had a week off a year, roughly speaking. He started work at fourteen and he quit when he was sixty-five, so he worked fifty years. Now multiply that out; he worked 150,000 hours in his lifetime. And then, without going through the math again, with our kind of work week and our kind of vacations and public holidays, and the fact that we don't go to work until we have a Ph.D., and the fact that we retire at sixty, or even fifty-eight, or whatever the unions want us to do, we will probably work sixty thousand hours, which is ninety thousand hours less than my grandfather worked. Now, that, sir and madam, is a revolution, and it cannot happen again. There are not ninety thousand hours more to chop off. The revolution has happened. The leisure problem is with us. We have solved it, you know; and how! We have solved it.

All right, let's look at some more pictures: Chart 3.

At the top left is what is called an arithmetic line. Here is the situation. The futurist plots ten points, years one through ten; and he sees that it makes a straight line. What are the chances that the next five years will do the same? And he says, being a sort of ordinary chap: "Well, I don't know anything to change it. Let's assume it will go on." Then you get an arithmetic forecast. This is the old style—you know, twenty million a year on population—and that's what happens.

CHART 3

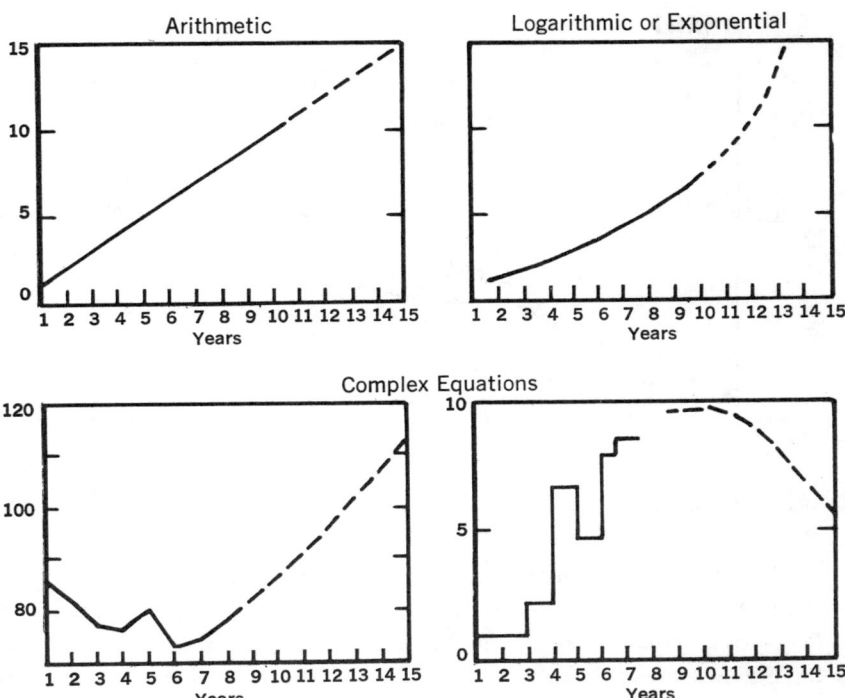

Suppose, on the other hand (top right), you get a curved line. This is what happens if something increases by 10, or some percent, steadily every year.

These sort of calculations the futurist is able to do in his head or on paper or with a slide rule. But for the graphs at the bottom, he has to use either equations of the order $x + y^2 + z^3$, or he goes to a computer, puts the ten-year data into it, tries to get a good fit, and asks the computer to forecast. And if you look at the bottom left, you can see there's something which is going down sort of hesitantly, and then up-down—and you think it might go up, and in fact the computer said it will. And on the bottom right we have one that's going up, but as the years pass, it's not quite so extremely up—and you have the feeling that maybe it's about to peak out, and it does. As a matter of interest, the first one is the pattern made by foreign aid in Canada. We gave foreign aid, then got very dissatisfied; then, I think, Lester Pearson said something and the government said oh, yes, and went back in line, and now we look as if we're going up. The bottom right is the interest rate on house mortgages. Okay.

CHART 4

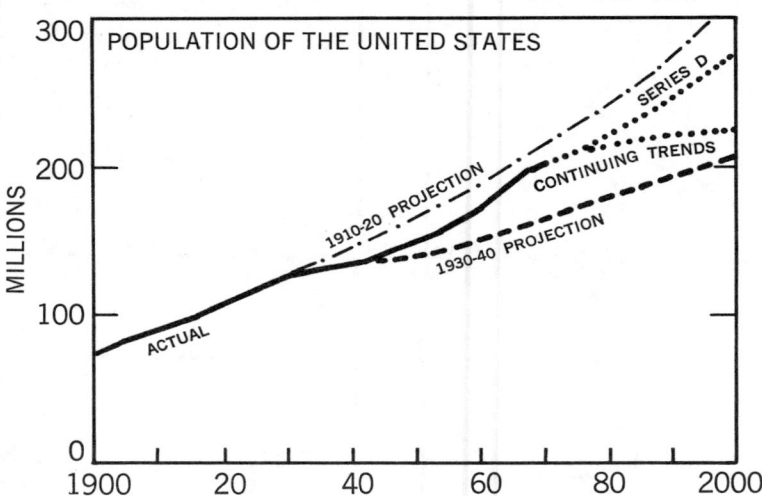

In Chart 4, the futurist is trying to understand what's going to happen to population in the United States. You'd expect population to be a geometric or exponential curve, because if people keep on having two and a half babies per family, pretty soon there are more families, and so there are more people who have two and a half babies. And this is true; population is a curving line.

Well, if we had been in the year 1920, and we had looked at the years 1910 to 1920 and said, that's the rate—all right, it's fairly well established—we would have drawn that line of dots and dashes, and you know, it wouldn't have been a bad fit. It would have been quite close in 1967. And if it went on up to the year 2000, it would be 330 million.

If we had done the same thing in the year 1940, the curve would take us up to a year 2000 population of about 209 million. Well, that's got to be wrong, because we're at 205 million right now—but I put it on for reasons which will become apparent. These two curves, in fact, bracket the present curve labeled Continuing Trends. Now, the curve labeled Series D is more complex. It's going up, but it's going up kind of decreasingly, you know—and that's possible, if zero population growth catches on, or if a whole lot of people go on the pill. That simply is an exact mathematical extension of the trends of 1965-1970, and it comes out at around 280 million. And the Series D curve is what the U.S. Census Bureau says might happen. It's one of four possible forecasts the bureau offers. The bureau has a series A, series B, series C and series D projection. Series A is absolutely ludicrous; nobody believes it any more, because it's so high. D is the lowest one, and they've just commissioned a lower one, which will be somewhere between the Continuing Trend line and the Series D line.

Well, the interesting thing is that with the sophisticated approach of the Census Bureau, which looks at infant mortality, and fertility in the age range 15 to 19, and length of time for a generation, which is changing, and so on—they come up with curves which are very close to my naïve approach. And so one can have some confidence in one's naïve mathematics. I urge you to do it, and have fun. And I just want to add that nothing I see looks very much like a population explosion. I don't know what it looks like to you, but very wise people keep telling us in the newspapers that the population of the United States is going to double by the end of the century. Well, you know, it's not. It's 205 million now, and there's no 400 million or 410 million on the chart.

4. *Don't Speculate When Good Facts Are Available*

You wouldn't have to speculate about the population of the United States, really, because the Census Bureau is doing it for you. Let's think about Picasso. Will he be alive in 1980? Picasso was born in October, 1881, and he's now eighty-nine. In 1980 he would be ninety-nine. Now, you can go to the insurance companies, and they have very good figures. You can say: What are the chances that a man of eighty-nine will live to be ninety-nine? And they'll say, all things being equal, the chances are forty to one against. There's the answer: "Improbable."

If Picasso had been born in 1891 the chances against his living till 1980 would be four to one, which is rather different.

If he'd been born in 1901, the odds would be fifty-fifty. If he'd been born in 1902, better than that.

5. *Distinguish Fashion and Fantasy from Real Desires*

Differentiate between serious trends and fashions and fantasies. People inject their fantasies into calculations of the future. They talk about domes over cities. They talk about immortality. They talk about robots. They talk about space travel. And yes, all right, that has happened, and is happening, but it is a fantasy, and the cost of it is almost the most fantastic thing about it. But everybody insisted that this one happen. It is essentially a fantasy.

Somebody was interviewing one of the pop singers in London, and asked him why he'd bought a Rolls Royce. You know, it's not in the image of pop singers. And he said: "Well, you feel such a fool without one." And this is a very interesting remark, because you sense instantly that the man is serious, and you can understand that the upper-income people think this way. I think it tells you something about affluence, and I've put it down as a thing I want to remember.

I noticed the other day that Canadian librarians are spending almost 20 percent of their operating budget on nonprint material. Librarians! The most conservative people, operating in that print world. That's not a fashion.

Let's look at this business of vacationing on the moon. So far it has cost something like twenty billion dollars to put four people on the moon and get them back. So the cost is on the order of five billion dollars.

Now, that's too much to spend on a vacation. But the airlines have been reducing seat mile costs about 40 percent per decade in recent years, and look as if they could go on doing so. Therefore if we are to say that the present cost of moon travel is about a thousand dollars a mile, and we want to get it down to one dollar a mile, we'll have to wait, at the present rate, 130 years, till the year 2100. It will still be $500,000 to go there and back, but in that year it may be not a bad price.

You could say that NASA will do better. You could say NASA's going to get this thing down in three decades, not thirteen decades—down to a dollar a passenger a mile. I still think that's not vacation money, you know—half a million dollars.

So I say the answer to that is "No." Politicians on junkets from Washington, yes; vacations, no.

When will surgeons transplant a human head? Now, this is not a fantasy. A lot of people are thinking about brain transplants; heart transplants are the reason why. Surgeons have said—surprisingly reputable men have said—that it will be easier to transplant a head than a brain. What makes it more difficult than transplanting a heart is connecting the many nerves to the many functioning parts of the body. However, a German medical team has done something about this; has managed to get nerves to grow along leaders, rather like vines growing on a piece of string. So let's say, technically the ability to do this will be here within a decade or fifteen years.

Now something more significant comes in, because we have to ask ourselves: What do we feel about head transplants?—or is it body transplants? You know, is it Smith if he's wearing Jones' head? Or is it Jones if he's wearing Smith's body? The thing is sort of ridiculous, in fact, it's repelling, really. It's a funny picture, but when you begin to think about it, you remember the military engineers in the war ripping up jeeps that had crashed and using the good parts, and so on, and the whole thing gets so mechanistic that you begin to wonder if anybody would want to do it all.

And so I would say you could answer this question in two ways. You could answer it on a technical basis; you could answer it on a social-emotional basis—and let's see.

6. *The World Is All One Piece*

Compare the part with the whole and the whole with the part. You know, this is microcosm—macrocosm. Look at what is happening in the corporation and ask if such things can spread into the world. We have better and better communications in business, very good Xerox machines, very good telephones, very good this, that and the other. Is a demand growing at home for such things? Look at what's happening in the world—for instance, the revolutionaries that are around: the FLQ in Quebec, the PLF in Palestine and the Tupamaros in Uruguay, and ask: "Can they come into the corporation? Is there something about the corporation that makes it different from the world? Why should somebody not hijack the president's Cadillac?" And ask the question seriously. Put the things backwards and forwards. The point is, the world is seamless; the world is one thing. We are not pigeonholed. Our activities are not separate in this way. We really are the same people when we're having breakfast, or when we're walking on the beach with our wives, or when we're at our desk, or when we're giving a talk, or when we're flying to New York, or whatever. If you have an idea, you might very well carry it somewhere else. So don't think of a revolution as being something that is guaranteed to take place anywhere but your office. Or don't think that a computer is something useful to the businessman, but housewives should beware of it.

Keep your ears open, and keep your imagination open for ideas that are becoming a little bit more than freakish.

There was an interesting interview in one of the news magazines with an Arab guerrilla girl. She said something like this: "The whole world ignored the Palestine problem; the deaths of Palestinian children, and so on, for twenty years. Now we are making you notice." The time of this interview was after the three jets were hijacked in the Middle East. "Now we are making you notice. If we throw bombs and so on, it is not our responsibility. We are not responsible."

You read the interview; you think: Well, who the hell is responsible? And, you see, the implication she is throwing is: You ignored us. Therefore, you are responsible. You made us do it. And that's a very interesting comment. Listen to that sort of thing.

Listen and notice when many things begin to tell you the same thing. You've looked at the statistics of industrial employment. In 1945 25 percent of the work force was in the service industries, 75 percent in the goods industries. In 1970 something like two-thirds work in the service industries. Massive switch.

See the kids walking down the street with a button on their lapel that says: People, Not Things. Look at Commander Bucher, who gave up the *Pueblo* so as not to risk the lives of his crewmen; who gave up the secrets of his country, the principles, if you like, of his country, gave up a whole lot with that ship and its intelligence material, in order to save eighty lives. Remember your father—your grandfather, maybe—fought in World War I, where they gave up many millions of lives in order to save things and principles. That's a very distinct change. Bucher is saying: People, not things. When they wanted to court-martial Bucher, most people in the United States said no. They said: People, not things. Very important—eighty lives. The service industries: People, not things. Most of us work with or for people, not things. Kids: People, not things. Listen to it. You can get these ideas. The welfare state comes out of that. Medicare comes out of that. The anti-technology movement, the ecology movement, conservation movement, and so on, come out of that.

7. *Read Science Fiction*

Listen to the science fiction people. Read science fiction. Here's

Arthur Clarke speaking about science fiction: "Over the last thirty years tens of thousand of stories have explored all the conceivable and most of the inconceivable possibilities of the future. There are few things that can happen that have not been described somewhere in books or magazines. The facts of the future can hardly be imagined *ab initio* by those who are unfamiliar with the fantasies of the past."

Clarke's record is very good. Clarke was talking about synchronous satellites in 1945. They happened in the sixties.

Science fiction people get it wrong sometimes. H. G. Wells said: "I do not think aeronautics will ever come into play as a serious modification of transport and communication." But Wells got a whole lot of things right. In 1901 he predicted a world war in 1940—very good. And in the 1930's he predicted the atom bomb. A lot of science fictioners were writing about the atom bomb in the 1940's when the Manhattan Project was going on, and General Groves wanted to have them stopped. The science fiction writers pointed out: "If you do, all the science fiction fans will know something's up." So the funny thing was that all the descriptions of the atom bomb were in the science fiction novels—worth listening to.

8. *The Conventional Wisdom*

Any trend that is apparent to everybody and agreed to by everybody is probably wrong. People started talking widely about the youth problem just at the point where youth started to be a declining proportion of the population. People started talking about the population explosion just at the point where population growth was going lower than it had ever gone before. I simply state those two things and ask you to think what they mean.

9. *The Lessons of History*

Remember that historians are experts, so be critical. The historians have told us that the cities are civilization and have then pointed at cities like Positano and Rome and Carcassonne as examples of what it's about. You also have to read history to know why cities got that way. They were planned cities. They were planned around planned societies, societies that were planned around the castle of the feudal lord, planned around church life, planned around the farm-market system, planned around whatever it was. The industrial revolution ruined that sort of planning. A planned city became impossible. I'm still against planners.

10. Integration

Do think about the quantitative measures of change, which I've discussed perhaps too much in this talk, but remember that they will usually be about economic or, at best, socioeconomic things. Recognize that *qualities* are not easy to measure and therefore not easy to discuss in that way, but are probably more important. I meant to get that point over.

What I think we are going into is a world increasingly shaped by consumers, not a world to be shaped by producers. That's a capsule remark which is so dense as to be almost meaningless, and if anyone wants to challenge it, I'd be delighted. But I hope I've hinted at that a bit. If I throw four balls in the air, the people-not-things thing; the generalized-not-specialized thing; involvement, not system; dynamic, not static, or, if you like, geometric, not arithmetic—if you connect all those things, and you then get what you think is a rather obvious picture of how the world is going to work, and you say to yourself: Well, that's a talk I could have given; I haven't learned anything new since I came here . . . then, I think, I will have succeeded. Thank you very much.

John Kettle

John Kettle is a journalist and writer well known in the United States and Canada as a leading futurist. In his books and articles he has presented and analyzed many of the problems that society must deal with in the remaining years of this century. He serves as consultant to the Hudson Institute, a leading "think tank," and is a member of the World Future Society in Washington, D. C.

Mr. Kettle was born and educated in England. After serving in the British army from 1946 to 1949, he joined the staff of *The Surrey Comet* newspaper as a feature writer and film and theater critic. In 1953 he emigrated to Canada, where he presently resides, and joined Hugh C. McLean Publications (now Southam Business Publications), where he worked until 1961 as an editor, first of construction magazines and then of *The Canadian Architect*, a magazine he founded. In 1961 he helped start *Canada Month,* of which he is still a vice president and director.

Since 1966 Mr. Kettle has worked as a free-lance writer for a number of publications. As special projects editor for *Monetary Times* he developed the "2000" series in 1967, the "Markets of the Seventies" series in 1968-1969, and the "Footnotes on the Future" series which began in 1970. This last is now published as a newsletter in *Executive* and *Canadian Doctor* magazines, of which he is special projects editor. He has contributed numerous articles to magazines and newspapers, including, among others, *Canada Month, Child and Man, Monetary Times,* the *Toledo Daily Star,* the *Toronto Telegraph,* and the *Times* of London. He has co-authored two books, served as series consultant to the CBC-TV series "Footnotes on the Future" in 1968-1969 and acted as consultant to and host of the 1970 CBS-TV "Landmark" series.

Mr. Kettle has also pursued an active interest in education. He was a contributor to the Greater Cleveland Social Science Program of the Education Research Council of Greater Cleveland (Ohio) and worked on an information project for the Centennial International Development Program of Ottawa in 1967. In 1968-1970 he was a director of the Toronto Waldorf School and served as president of the Waldorf School Association of Ontario, Inc.

BIBLIOGRAPHY

and Walker, Dean, eds. *Verdict!* New York: McGraw-Hill, Inc., 1968.

and Walker, Dean. *The Awkward Stage.* Toronto: Methuen, 1969.

Footnotes on the Future. Toronto: Methuen, 1970.

ed. *Beyond Habitat.* By Moshe Safdie. Montreal: Tundra, 1970, and Cambridge, Mass.: The M.I.T. Press, 1970.

8. FORCES FOR CHANGE IN THE FINAL THIRD OF THE TWENTIETH CENTURY

HERMAN KAHN
Director
Hudson Institute

We have a great deal of ground to cover so let's plunge right into the subject. Can everybody read Chart 1? It is what we call an amilitary-apolitical surprise-free projection. The jargon is fairly important. The amilitary and apolitical are telling you that nothing is happening—which is right for this projection. One of the basic ideas here is that you could go to the drugstore today and buy yourself a map of the world and, except for maybe parts of Africa and Asia, it would hold pretty well for the year 2000. That would not have been true had you bought the map in 1900 or 1933.

As for surprise-free projection, in this kind of business we talk about three things: projections, forecasts and predictions. A projection is simply a method of extrapolating into the future; no assertion of validity. A forecast tries to describe, say, the horses that are running and their likely odds. If I have a coin, and it's a fair coin, my forecast is 50 percent heads, 50 percent tails. A prediction would say: Call it heads.

Now, a surprise-free projection is a special kind of projection. It says: Put in all the theory that you believe. If it then turns out to be true, you shouldn't be surprised at your theory. If you have two theories, and they are contradictory, that's all right too; if you aren't sure which one is right, you still won't be surprised if one or the other comes out.

Most Hudson staff members at this point make a comment about this kind of projection: that the most surprising thing which could happen would be to get no surprises; and therefore you shouldn't take the projection too seriously. You see, the projection is a context with which one can disagree as well as agree. It's simply laid out to make the kinds of things we're talking about very specific; but there is no assertion that we know what is going to happen.

That's a disingenuous way to talk, and it makes friends. But I want to take the other position. This is a prediction. I've been looking at Chart 1 for four or five years, and I think it's right. And the analogy I'd like to draw is with the period 1815 to 1914.

CHART 1 **Relatively amilitary, relatively apolitical, "surprise-free" projections of the most signficant aspects of 1970-2000**

1. Continuation and/or topping out of multifold trend.
2. Onset of post-industrial culture in nations with 20% of world population, and in enclaves elsewhere.
3. "Political settlement" of World War II—including the rise of Japan as the third superpower (or near superpower) and the reemergence of Germany.
4. With important exceptions, erosion of the 12 traditional societal levers and a search for meaning and purpose.
5. The 1985 technological crisis: need for world-wide (but probably ad hoc) "zoning ordinances" and other controls; possible forced topping out of no. 1 above.
6. Onset and impact of new political milieu.
7. "Humanist left" vs. "responsible center" confrontation—particularly in the high (visible) culture.
8. Increasingly "revisionist" communism, capitalism, and Christianity in Europe and Western Hemisphere.
9. Decrease in consensus and authority; increased diversity (and some increased polarization) in ideology, value systems, life styles.
10. Increasing problem of trained incapacity and/or illusioned or irrelevant argumentation.
11. World-wide (foreign and domestic) law-and-order issues.
12. Populist and/or "conservative" backlash and revolts.
13. Better theory and application of methods for sustained economic development almost everywhere.
14. High (1-15%) annual growth in GNP/capita almost everywhere.
15. World-wide capability in industry and technology; growth in multinational corporations and conglomerates.
16. Turmoil in Afro-Asia and perhaps Latin America.
17. Nativist, messianic or other "irrationally" emotional mass movements —general decrease in rational politics.
18. Relatively multipolar, orderly and unified world: enormous growth in trade, communications, travel; limited development of international and multinational institutions; some relative decline in power and prestige of U.S. and U.S.S.R.; emergence of "intermediate" powers (East Germany, Brazil, Mexico, Indonesia, Egypt, Argentina, etc.); possible bid by Japan for world leadership of some sort; China and Europe both rise and fall.

If you were at the Congress of Vienna in 1815 and asked the people there, "What do you think is going to happen?" almost unanimously they would have said, "Twenty-five more years of revolutionary violence, just like the last twenty-five." What they got was a hundred years of relative peace. They had 1830, 1848, the Crimean War, the Austro-Prussian War and the Franco-Prussian War—but these were trivial compared with what they expected. Mostly, that period was taken up with evolutionary change.

What I'm suggesting is that today we're in for an evolutionary period rather than for the kind of violence we've become used to in the last seventy years.

The first three items on Chart 1 also have rather nice analogies with 1815. The most important thing that happened in the nineteenth century was the industrialization of Europe. That corresponds to the multifold trend, which is a trend in man's affairs, in Western culture, which goes back not quite a thousand years. We'll look at a chart in a moment that shows what I mean by the multifold trend. But you can think of it as something which goes back, in some of its parts, two or three thousand years; in some parts, six hundred years; in most parts, nine hundred years.

THE NEXT THIRTY YEARS

The most basic assumption of the study is that what's been going on for nine hundred years will continue for the next thirty. That takes more courage than you might think; there are a lot of ebbs and flows in this trend, you know, but I am arguing that the trend will flow, so to speak; not ebb. In a longer discussion I'd give you a lot of scenarios which would make it ebb.

The next thing that happens is development of the post-industrial culture. I'm going to make a very big claim. When I discuss this post-industrial culture I'm going to use the most pretentious language of which I am capable. I'm going to argue that the next ten, twenty, thirty years will see as big a change in the history of mankind as has ever occurred before; and I'm going to use pretentious language to drive that point home. And by the way, I'm capable of very pretentious language.

The third point is, I think, extraordinarily interesting, again by analogy. The one thing which nobody expected in 1815 was the rise of Prussia. The general expectation was that a five-hundred-year war between England and France would more or less continue. And Russia and the United States loomed in the background. Tocqueville made a famous prediction in 1830 which was widely believed: that eventually the world struggle would be taken over by Russia and the United States, that their confrontation would replace the cold war, if you will, which had always existed between England and France. He said that an equal rivalry would emerge between Russia and the United States—which was not a bad prediction. But it's a prediction that would not have worked out if the Kaiser had won his war or Hitler his war, other things being equal. And nobody in 1815 had any idea that the really big issues of the late nineteenth and early twentieth centuries would revolve around Prussia—it was totally unexpected.

I'm going to suggest that the same thing may be occurring today: that the big issues of the future are going to revolve around the rise of Japan. We're doing a study right now for some thirty United States and thirty foreign corporations, and the main point we have to spell out for about half of these people is that they are going to have to adjust to Japan; and the adjustment is going to be phenomenal. You don't make tape recorders in this country any more, you don't make radios. By 1980 you won't be making TV's in this country; you won't be making phonographs—and so on. I don't know much about Western Electric manufacturing, but I'll make a bet that half or two-thirds of Western Electric's products are no longer manufactured in the United States ten years from now.

What's going on here? Well, in the fifties the Japanese economy grew by a factor of two from small to medium. That was rather impressive—but nobody except the Japanese really cared.

In the late 1960's it grew again by a factor of three. That's unprecedented; and on top of the first factor of two, it's extremely impressive. Japan grew from medium to large, becoming the third-largest economy of the world. By now it's around 200 billion, or about 20 percent of the U.S. economy and about one-third the Soviet economy.

It is now believed—I don't know of anybody who doesn't agree—that the Japanese economy ought to do better than grow by a factor of three in the seventies. Some people talk about four, some talk about five—I don't know—give it three. This still means that it will grow from large to gigantic. It will become very noticeable. When an economy moves from large to gigantic, it makes waves in a way that it does not in the first two phases.

The fourth item in Chart 1 is an interesting set of ideas which have to do with the changes in our own culture: if you will, the hippie phenomena, the New Left phenomena, the protest phenomena, the student issues. The backlash—the so-called backlash—also comes in here. What's going on in the value systems in America?

Finally, I'll discuss today the 1985 technological crisis. This is a peculiar phenomenon. The term comes from a 1955 article by John Von Neumann in *Fortune,* where he suggested that we would be running out of room around 1980. All kinds of processes which seem to operate fine now will suddenly come to a stop. If you want to be dramatic and very pessimistic, it's as if you put up a Hollywood set and fixed it to last till 1980. You expect to use it until then and afterwards throw it away. So you won't mind that in 1981 the plumbing breaks down, in 1982 the electricity breaks down, in 1983 the fire hazard becomes enormous. These make no difference because the thing will be abandoned in 1980.

Now, that's not exactly the prediction I want to make about the world. But the impression I do want to give is that an unbelievable number of problems will come to a head around the mid-eighties, problems which we have been ignoring.

It happens that the remark I just made has become a fashionable one. I remember giving a talk in 1957 and discussing one of these problems—pollution—and suggesting that it's difficult to exaggerate the pollution problem, but that we'll probably succeed in doing so when it gets fashionable—and we have. If you go down to the magazine stand today, every other magazine has an article on "We're Drowning in Our Own Garbage"—*Screen Romances, Cosmopolitan, Ladies Home Journal*—every serious or not-so-serious magazine of opinion in America is now turning its attention to this issue.

I will try not to add to the kind of overstatement you get now. I do want to make clear that the problem is important, does exist, is quite dramatic—and is currently being overstated, as you might expect.

MULTIFOLD TREND

Let us turn now to the multifold trend, outlined in Chart 2. I claim that this trend has been going on for about nine hundred years. A lot of people have noticed it, so there's almost nothing on the chart that's new or interesting; if there were, it would not belong here. Some of the items, however, deserve comment. The trend towards an increasingly sensate—this-worldly or secular—society is probably the most important.

The trend has developed to such an unbelievable extent that most people in the world—not just Americans—have little or no concept of what sacred means. We have so desacralized our culture that the mere concept of the authority of the sacred is far removed; you've got to explain it to people. I'm not going to have time to do that today. But I have some charts which try to do it. (Charts 17 to 20 at the end of this article.)

CHART 2   There is a basic, long-term, multifold trend toward:

1. Increasingly sensate (empirical, this-worldly, secular, humanistic, pragmatic, manipulative, explicitly rational, utilitarian, contractual, Epicurean, hedonistic, etc.) culture. Recently, an almost complete decline of the sacred and a relative erosion of "irrational" taboos, totems, charismas.
2. Bourgeois, bureaucratic, "meritocratic" elites.
3. Accumulation of scientific and technological knowledge.
4. Institutionalization of technological change, especially research, development, innovation, diffusion. Recently and increasingly, a conscious emphasis on synergisms and serendipities.
5. World-wide industrialization and modernization.
6. Increasing capability for mass destruction.
7. Increasing affluence and (recently) leisure.
8. Population growth—now explosive, but tapering off.
9. Urbanization and (recently) suburbanization and "urban sprawl." Soon, the growth of megalopolises.
10. Recently and increasingly, macro-environmental issues (constraints set by finite size of earth and various local and global reservoirs).
11. Decreasing importance of primary and (recently) secondary and tertiary occupations.
12. Increasing literacy and education. Recently, growth of the "knowledge industry." Increasing numbers and role of intellectuals.
13. Future-oriented thinking, discussion, planning. Recently, some improvement in methodologies and tools—also some retrogression.
14. Innovative and manipulative rationality increasingly applied to social, political, cultural and economic worlds, as well as to shaping and exploiting the material world. Increasing problem of ritualistic, incomplete, or pseudo-rationality.
15. Increasing universality of the multifold trend.
16. Increasing tempo of change in all the above.

I had a grandfather who used, literally, to talk with God. He got his messages early in the morning, carried them out during the day and reported at night to check that everything was going according to plan.

You know the recent papal encyclical on birth control? A number of people said—in fact, there was almost a world-wide protest—that this could not be good church dogma because it would increase human unhappiness. My grandfather would have said: "What else is new?" The mere concept that you judge religious dogma by referring to human happiness would be foreign to him. And in fact, the idea is only three or four hundred years old. Nobody in the fifth or the tenth century ever used an argument of that sort; no scrap of paper has ever come down from the fifth or the tenth century which mentions human happiness. It was not an issue of importance to the people who wrote in that culture. And when those people talked about the sacred, they weren't talking about what we are today, which is typically religious humanism.

I spend a lot of time with young Jesuits, and I don't know of any who can be called Catholics as the term was used ten years ago. And I make that statement before Jesuit organizations, and nobody argues. They are religious humanists, and they would deny that this is being irreligious. But they are not Catholics as the term was used ten years ago.

Now, that's true of the other religions as well. In the 1900's we saw in this country the atheistic-agnostic-theist minister. In the twenties and thirties we saw the atheistic-agnostic-theist rabbi. We now see the atheistic-agnostic-theist priest. And it's not just Jesuits. You find it in all the church orders.

I just spent a morning talking to a school advisor—my daughter is graduating from high school—about what kind of college to send her to. My wife suggested we might send her to a convent; and the school advisor said, "No, no, they're swinging now too."

To really finish this point, I was down at Texas A & M last week. And it was kind of pleasing—I'm a relatively square bourgeois type, and there were all these nice, clean-cut kids, and the girls all dressed to the hilt. I thought, gee, that's where Debbie ought to go. I suggested that, and the newspaper picked it up. The next day I was talking at Southern Methodist University, and somebody in the audience suggested, "Why don't you send her over here instead of to Texas A &M?" I looked around and said, "Too many beards." It's a square school, but my standards are very high. Anyway, just to get all the prejudices out—we now live in a secular, humanistic, pragmatic, manipulative, rationalist, utilitarian, contractual, epicurean, hedonistic, etc., culture.

POST-INDUSTRIAL SOCIETY

I will make a few comments on the post-industrial culture because that's what we're most interested in today. (You can see its development on Chart 3.) First, a few definitions: Primary occupations are farming, fishing, agriculture, forestry, mining—you know, occupations in which you get your living right from the earth itself. So it's no problem explaining to your two-year-old son how you're making a living. It's clear and understandable; he likes to emulate your work.

Secondary occupations are in manufacturing and construction. Again, they are concrete operations, and your son can understand what you're doing for the world.

Tertiary occupations are a little tricky. They're services: communications, telephone, transportation, insurance, finance, advertising, marketing—involving roughly every man in this room. With these it's very hard to explain to your son what you're doing.

And what's happening today is that there's been a steady decline in the importance of primary, and more recently of secondary and tertiary, occupations.

What's left? Something which I will call quaternary. Quaternary occupations are peculiar. They're service occupations, basically, but not services which help production. For example, say people do a lot of skiing; that's done for its own sake, not for production. The man who runs the hotel where you ski is giving you a service, but it's quaternary, not tertiary. If you're a poet and somebody teaches you English, that's a service occupation, but quaternary, because the poetry is not oriented towards social production. If you're taking photographs for their own sake, any services like developing, and so on, are quaternary occupations.

Increasingly we're getting into what's called a quaternary culture. There's some importance in this. We're going to deal with future-oriented thinking. We're trying more and more to control our environment, partly because we have to; that is, if technology and affluence create pollution, somehow more technology and more affluence may be the solution, but intelligently applied.

Some of our concern has to do with just the magnitude of the deci-

CHART 3 One way to look at man's economic progress

ANNUAL PER CAPITA PRODUCT (1965 $)	ECONOMIC SYSTEM	MOST OF DEVELOPMENT
0	Pre-agricultural or primitive	1st 500,000 to 2,000,000 years
$50–$250	Pre-industrial or agricultural	8th–1st millennia B.C.
$100–$300	English industrial revolution	1760–1790
$200–$1,000	World-wide industrialization	19th–21st centuries
$500–$2,500	Mature industrial (Western Europe)	Mid-20th century
$1,000–$10,000	Mass consumption	Mid-20th to 21st century
$5,000–$25,000	Post-industrial	21st century
$50,000–$250,000	Almost post-economic	22nd century (if annual GNP/capita increase is 2.3% or so)
$500,000–$2,500,000	?	23rd century

sions. If you're a President of the United States, you don't like to admit that major changes in policy were made off the top of the head. Things like the Marshall Plan, the Truman Doctrine, Point Four— none of these had been studied before they were announced. You know, they were created in response to the need for certain kinds of speeches. Very hard to admit that. Back in the Roman Empire, we could have read the entrails of chickens and let them point out that this was the right time for a certain change in policy. We don't read entrails any more, but you can get studies, and many of them are not too different from reading entrails. I wouldn't knock the entrails-type study; it's a lot more useful than you think.

In fact, I had a very upsetting experience, I guess two weeks ago. I had a birthday and one of the girls at the office put out ten dollars for a computerized horoscope—right on the button, sixteen pages that described everything about me to a tee, and all of it good, of course. But it's kind of impressive.

All right, anyway, technological development and the tendency toward quaternary culture is the basic trend of Western culture for the last thousand years. I want to look at this trend from the economic point of view, and I want to suggest that it may be topping out, or changing its character in a radical fashion.

Now, I said that when I raised the issue of the post-industrial culture I would use the most pretentious language of which I am capable. One way to do this is to use numbers Say, refer to the fact that there are a hundred billion stars in the galaxy. Somehow anybody who thinks about that number of anything must be a deep, wise thinker. If you say there are a hundred billion galaxies in the universe —you know, my God, that's unbelievable! Or if you say that man has been on earth for one or two million years, that's historical perspective. You see?

I want to use exactly that trick. Man has been on earth for one or two million years. I've examined every one of those years rather carefully. I have noticed only two events of any interest; the rest is trivial— an unbelievable amount of trivia, by the way. If you're a religious person, you really have to add a third event, and then again we might disagree on what the third event is. I refer, of course, to the covenant of God with Abraham; I gather that some of you would not pick that one as the third big event. Let's not get into an argument about it, and let's stay within the two events we can agree on— two secular events.

What were they? First, the invention of agriculture, about 10,000 years ago in the Fertile Crescent. That created civilization. What do we mean by civilization? We mean a civic culture, people living in cities. And in some sense the Chinese are just as civilized as the Americans, maybe more so. If the Indians have a problem, it's because they're overcivilized rather than undercivilized. To lump India and Africa together as underdeveloped countries is a peculiar thing. One has a 3,000-year civilization; the other is tribal—it's not civilized, if you will.

It's rather interesting that agriculture did not change the average standard of living. Roughly speaking, no agricultural system that we know of ever got much over $250 per capita, or dropped much under $50 (Chart 3). If you want to ask what is normalcy for human beings, it's India today, Indonesia today, China today—they get $100 per capita; roughly in the center of that range, you see.

Nothing happened for around eight thousand years, almost ten thousand years, and then came the so-called industrial revolution. In fact, as you would guess, that revolution can be traced back to Greek times. These things get spread out when you look at them more carefully. But a lot of things happened in the thirty years that we call the industrial revolution. An Indian who visited England in 1750 would have seen a country not very different from India; but in the year 1800 it was very different. There were factories everywhere, industrialized cities. And the standard of living did rise. But the crucial issue is not the per capita income. It's the fact that there was a completely different style of life and way of looking at things. Per capita income is not unimportant, however, and industrialism increased it by ten times; it made people rich.

We are now entering what I and many others call the post-industrial culture. In this category the annual income begins at about $5,000 per capita, which is roughly the per capita income in America today. Again, the crucial issue is not income, but the fact that the income is increasingly from quaternary rather than primary, secondary or tertiary work. I'll come back to that in just a moment.

Chart 3 includes an economic system which is termed "almost post-economic." There are about three hundred people around the world who write and study about such a system, and approximately two hundred of them, it seems to me, have it wrong. They call it post-industrial, but if they're talking about anything that's realistic at all, it's about the twenty-second century, not the twenty-first. If you continue growth rates at 2.3 percent a year, you should get another factor of a hundred in growth, which is not unreasonable. Much of what they say about the post-industrial society corresponds to a society with a per capita income in the $500,000 to $2,500,000 range. I use the term "almost post-economic" for that society. You never really get rid of the economic problem. I can imagine a family in the year 2200 that desperately wants to spend its vacation on Jupiter, can't raise the $5 million it takes, settles for a cheap $1 million vacation on Mars; feels desperately deprived—everybody else went to Jupiter. We don't sympathize. But these people would have the same kind of trouble sympathizing with us. They would not understand our economic problems; we would not understand theirs.

Chart 4 is a simple diagram indicating four stages of human history and three important transition points. Note that the single line designating the agricultural revolution took eight thousand years. It took eight thousand years for agriculture to reach Spain and England, so it really is a complicated process.

Very likely, if somebody were giving this talk, say, five thousand years from now, he wouldn't make distinctions involving only 250 years; in fact, he'd lump agricultural and industrial society together in one line.

CHART 4 Today we tend to divide man's economic history into four basic stages.

1st 2,000,000 years	Hunting and food gathering (Pre-agricultural, basically primitive)
8000 B.C. to A.D. 1750	Basically agricultural (Pre-industrial, sometimes civilized)
1750 to 2000	Industrial (Or modern and/or scientific) Industrial revolution Partially industrialized Mature industrial Mass consumption
2000 +	Post-industrial (Or post-modern and/or post-scientific) Emergent Visible (Mature) (Late)

Okay, why don't we try that? I suspect that what we sometimes call the big picture may look like Chart 5. This is man's history: pre-civilized for two million years; primitive, tribal. Civilized for ten thousand years; then something new. What is it? Well, I don't know. Nobody knows. Post-civilized? Post-economic? Maybe. Truly human? I don't know what that means. Post-human—you know, computer-man, made in the laboratory? Maybe. Faustian—you know, where he's really sold his soul to the devil? Remember, there are two Fausts: Marlowe's, where the devil collects, and Goethe's, where Faust goes to heaven. But both sold their souls for profane knowledge, profane power. Post-Faustian? Promethean? Godlike? You know the deist position: they believe there is a God, a first cause, a supernatural, transcendental being; but they don't claim to have any information about it. It's sort of like the Unitarian position—there is at most one God. The deists may be talking about what man will become.

What I'd like to do now, for the rest of this hour, is move in closer on this big picture and make a few specific predictions.

Chart 6 gives a basic picture of the post-industrial culture as we see it today. Notice in the heading the adjective "emergent"—we don't know what's going on: we're like Dickens in nineteenth-century England. He had a very strange picture of an industrial society, but he knew, roughly, what would be going on in England in the year 1900. We know, roughly, what will be going on in the United States in the year 2000.

CHART 5 Future man may use only three stages.

1st 2,000,000 years: Pre-civilized

8000 B.C. to A.D. 2000: Civilized

2000 to 2010? 2100? Eternity? Until fulfillment?

        Post-civilized
        Post-economic
        Truly human
        Post-human
        Faustian
        Post-Faustian
        Promethean
        Post-Promethean
        Godlike
        Truly religious (neo-deist)

I've already mentioned quaternary activities, income growth by a factor of 100 more than in a pre-industrial system, and the deemphasizing of narrow economic efficiency. Now look at Chart 7.

A Post-Business Society?

Before World War I, or before 1930, when an American was applying for a job he really asked only two questions: What is the salary? What are the chances for advancement? During the depression of the 1930's, it was difficult to move Britishers out of the so-called depressed

---

CHART 6 **The emergent, U.S., year 2000, post-industrial or post-mass-consumption society**

1. Most economic activities are quaternary (self-serving, or services to self-serving activities, or services to such services) rather than primary, secondary or tertiary (production-oriented).
2. Per capita income $5,000 to $25,000/year (or about 10 times that in industrial system and 100 times that in pre-industrial).
3. Narrow economic "efficiency" no longer primary.
4. Market may play diminished role compared to public sector and "social accounts."
5. Official floor on income and welfare for "deserving poor"; effective floor for others.
6. Perhaps more "consentive" and anarchic-type organizations (vs. "marketives" and "command systems").
7. Business firms may no longer be major sources of innovation or centers of attention.
8. Widespread use of automation, computers, cybernation.
9. Small-world "global metropolis," not a "global village."
10. Typical "doctrinal lifetime" 2 to 20 years.
11. A "learning" society—emphasis on late knowledge, imagination, courage, innovation. Deemphasis on experience, judgment, caution—perhaps wisdom.
12. Rapid improvement in institutions and techniques for training and teaching; "education" may lag.
13. Erosion (in some upper and upper-middle classes) of deferred-gratification and work-, achievement-, advancement-oriented values.
14. Likely erosion (at least in some upper and upper-middle classes) of the other 11 "traditional levers."
15. Much apparent "late sensate" chaos and polarization.
16. Sensate, secular, humanist, perhaps self-indulgent criteria may become central in important groups—at least during this transitional period.
17. But the search for "meaning and purpose" will largely find at least an interim solution (or solutions).
18. Important elements of this solution may be "against progress," against nos. 15 and 16 above, and/or against "Western culture."

CHART 7 Before World War 1, the average middle-class American asked two questions when applying for a job:

1. WHAT IS THE SALARY?
2. WHAT ARE THE CHANCES FOR ADVANCEMENT?

areas. They wouldn't move five miles to a job; they wanted to stick very close to their home base. That's the way human beings have traditionally been; they are reluctant to leave the place where they were born. But Americans were different. We were very mobile, and this was one of the main reasons for our economic success. You could move a man from Maine to California for five cents an hour more. You couldn't move him back, but that was not an issue.

Today it's different; today an American will usually ask all the questions listed in Chart 8. He will still ask: What is the salary? What are the chances for advancement? But he will also ask: What are the schools like? What does it mean for my family? The old questions are not dominant anymore.

One way to see what's happening—and in general, a good method if you ever want to know the true heart and soul of a people—is to look at the third-class literature: at the soap opera, the confession magazine, the grade C movie. This kind of stuff speaks from the cliché heart to the cliché heart. It's not confused by either genius or creativity. It's genuine in a way that the grade A movie is not.

In grade C literature during the 1930's, whenever there was a conflict between job and family, between advancement and friendship,

CHART 8 Now he tends to ask:

1. WHAT IS THE NEIGHBORHOOD (OR WAY OF LIFE) LIKE?
2. HOW ARE THE SCHOOLS?
3. IN GENERAL, WHAT WILL THE IMPACT BE ON MY FAMILY?

He may also ask:
4. WHAT ARE THE FRINGE BENEFITS?
5. WHAT DO YOU DO (OR MAKE)?
6. WHAT KIND OF ORGANIZATION ARE YOU?

Before he asks:
7. WHAT IS THE SALARY?
8. WHAT ARE THE CHANCES FOR ADVANCEMENT?

But he still usually asks the last two questions.

job and advancement won out, or there was tragedy. The American who put family or friendship first created a tragic situation.

In grade C literature during the 1960's, unless the job involved something altruistic, like a psychiatrist's or a doctor's or a Hudson Institute director's, family and friends won out, or there was tragedy.

In 1930 soap operas, an American who earned a million dollars and picked up an ulcer in the process was looked upon as a hero, honorably wounded in the battle for success.

In the 1960's that same man was portrayed as a compulsive neurotic with twisted values. He's sick; send him to the hospital!

That's about as big a change as you can get in a thirty-year period. It's one of the big things that has already happened to our country. You may think it's healthy or unhealthy; I'm simply putting it forth as an observation.

As Chart 6 says, one important change in the post-industrial society is the change in the position of business itself. When we say post-industrial, we mean literally that industry no longer plays a central role in the economy.

How many of you know the book *The New Industrial State,* by Galbraith? All right. I'm taking the exact opposite thesis here. Imagine a man, say, a hundred years ago, when half of America was a farming society. Because he didn't know that in 1970 less than 3 percent of American farmers would produce more than 95 percent of the foods and fibers, he might talk about the "new agricultural state." Like Galbraith analyzing industry, he might argue that agriculture was going to dominate everything because of its tight monopoly on the most basic processes.

But today farming is not a dominant activity in the United States; it's considered a rather dull occupation. I can imagine a farmer coming home and telling his wife: You know, I've just doubled production in the south hundred acres; his wife saying: Gee, that's interesting; what else have you done today? You know, she's not going to be turned on. The *New York Times* may publish it on page 8. If you're interested in the food business, you'll read it; if you're not, you won't. It's not an exciting event in America.

Well, industry is going to be like that in the post-industrial culture. The very success of industry will make it dull.

Sometimes businessmen ask me: "Who's going to pay for this fancy state of yours?" Businessmen. Well, who's going to feed you? Farmers. Anybody here feel a recent pang of gratitude towards the farmers of America? Raise your hand if you felt a recent pang of gratitude. You know, they feed us without working very hard at it, as far as we can tell. And industry does its job without working very hard at it. There isn't much interest in either activity.

Will the post-industrial society be post-business as well? Yes, business stays in the same channels that it has occupied historically. If business moves into new areas—education, medical care, urban renewal, old-age care, maybe certain kinds of services — the society may well be post-industrial but not post-business.

There seems to be no sign in Europe of such a movement. There are definite signs of it in America, so I wouldn't be at all surprised if society became post-business in Europe but not post-business in America. You might have a different situation in the two areas.

I mentioned already that there are about 300 people who study this kind of thing. About 290 of them are convinced that society will become post-business. If you ask them where the center of activity will be, their answer is always: the university. These 290 are all college professors. But 10 of us think that the center of interest may be the policy research organizations. I won't mention who we are.

A Computer Society?

Let's look at some other points, starting with the computer, because it typifies these changes.

Most of you who are used to reading about computers or talking about them know what is generally referred to today as the fourth-generation computer. We had back-up tubes, transistors, solid state integrated circuits, and now you get large-scale integration—you know, fourth-generation.

Now, that's true of the switching equipment, if you will, but there are other criteria. Every 2.3 years since the late forties has seen an increase in the power of computers by a factor of about 10, and it's a nice straight line to chart; it's really kind of beautiful. We drew this curve about six years ago and the IBM 360 fell right on it—the new model. Until three or four years ago, nobody successfully got off that straight line of development. Two computers were built that were not on the line—the IBM Stretch and the Philco 2000—and they were both failures. The straight line was a better predictor of computer development than the American engineering profession.

I don't want to get into technological forecasting here. But there's no tendency for this line to top out, despite a lot of stuff you read to the contrary. That means you can expect, every two or three years, the power of computers to increase by a factor of about 10.

Now, what does this mean? Well, you've had 10 powers already in the last twenty years; that puts you in the tenth generation, because the power of 10 is a big enough change to make any ideas you had obsolete. So anything you learned three years ago can be wrong today about the advanced computers. Anything you're learning today may be wrong in three years about the advanced computers.

This is a business which contrasts with civil engineering, where the sixty-five-year-old man earns more than the fifty-five; fifty-five more than forty-five; forty-five more than thirty-five; thirty-five more than twenty-five. It's not a particularly intellectual society; it's one in which experience, grey hairs, judgment all count.

Computer technology? Peak salary is reached in your early thirties —thirty-one, thirty-two, thirty-three. After that you go back to school, or you join management and give up earning an honest living, or you accept the flattening out of your salary.

And the basic idea is that more and more of the world will look like computer engineering; less and less like civil engineering. There's going to be a shift of power, or authority, or center of action, from the old to the young. There's always a problem of generation change, you know. In some sense there's always a struggle for power in every society between the old and the young; it may be a central struggle or a peripheral one. What I'm suggesting here is that there will be a shift of emphasis, authority and responsibility towards up-to-date knowledge, courage, creativity and imagination; and a de-emphasizing of judgment, experience and maturity. This shift then changes other things, you understand. Of course, you might have the shift in computer engineering and in nothing else, but I don't think so.

It is almost impossible to get relevant information about the use of computers in business today. Some of you may know that a lot of financial companies, insurance companies, and banks went into computers in a big way in the mid-fifties. As far as I know — with the possible exception of the Bank of America—they all lost money. I say "possible exception" because there used to be a theory that the Bank of America had not, but I recently talked with some of its management people, and they said: "What makes you think we were an exception? You're not a person who normally believes an organization's publicity; why do you believe ours?" Let's count them as a possible exception. Everybody else lost money.

Then a lot of these companies lost money a second time because they failed to get into the computer business in the early sixties, when it was essential to be competitive. Having been bitten once, they were twice shy.

How could they make this kind of mistake? Well, it's still going on. I was in Europe last year at a very senior management conference, where some of the top people in America were talking to the top people in Europe about the technology gap. And a number of speakers made the same point: that in America we've got big companies with large integrated bases, with an integrated management information system, and unless the Europeans get on the bandwagon very fast, they're going to be hopelessly outclassed.

Now, it happens that there is no big company in America with a large data base of that sort; there is no big company in America with an integrated management information system. There may be some small companies, I don't know; but I'm taking about big ones.

These people didn't think they were lying; they're enthusiasts. And this is the problem you get into. The people who know about the new systems are almost uniformly enthusiasts; otherwise they wouldn't have bothered learning. The people who take a dim view tend to be totally ignorant, so you can't trust them either. You find in these fields that it is very difficult to get advice which is both mature and sophisticated.

To continue with changes in the computer field, let's assume that the factors of 10 in improvement go on for the rest of the century, resulting in a factor of something like 100 billion to 100 quadrillion. It's very hard to estimate what that means in terms of actual capability. It wouldn't be surprising if in the year 2000 this type of lecture was given by a computer to a bunch of computers. This might be the way they learn.

You may believe that there are intrinsic limits to computers. If you do, you arrived at that conclusion by revelation or intuition, not by rigorous argument. There has been no paper ever written which reliably demonstrates that computers have limits of any sort. I rather suspect that they don't, and if they don't, that's a very depressing thought, in my judgment.

It won't at all surprise me if at the end of this century computers can transcend human beings in every aspect of human life, including poetry, making love—I'll come to that in a few moments—painting, management—you name it.

They may not, of course. We just don't know.

Now, if there is some human activity which computers cannot perform, say playing tiddlywinks, that will become the major activity of mankind—whatever it is, that's how you're going to express your humanity.

Computers are built today, as you may know, that can probably beat every man in this room in chess or checkers. They can't beat a master, but neither can you. They can beat any average player, which means they can beat the players who built them; they can transcend their own makers.

This process is a part of the multifold trend. You know what the word "mediterranean" means? It means "the sea in the middle of land," the center of the earth. Well, the Greeks were wrong about that. Then people had the idea, remember, that the earth was the center of the universe—the Ptolemaic system. But it turns out that ours is only a small planet at the edge of a minor star, at the edge of a minor galaxy. And then we got the idea from Darwin that men and apes are descended from a common ancestor who looks more like an ape than a man. And then we got the idea from Marx that most of our motivations are economic; not really thought through, if you will. And then we got the idea from Freud that most of our motives are unconscious and, generally speaking, unworthy of us.

This, then, may turn out to be the final dethronement of mankind—that we can be beaten by a bunch of transistors.

To continue along another line: "upper middle class" is not so much an income definition as a syndrome of attitudes. Every Texas millionaire I've met is lower-middle-class, without exception. Most people in this room probably have a lot of lower-middle-class attitudes: ideas that are square, bourgeois—that kind of thing.

If you divide classes by income, you normally say that five to ten thousand a year in the North is lower-middle-class; ten to forty or fifty is upper-middle-class. But, I repeat, my Texas millionaires are lower-middle-class. A Jewish schoolteacher earning six thousand a year in New York is upper-middle-class in the basic attitudes we're talking about. If you don't like the terms "upper and lower middle class," you can think of "traditional America" and "progressive America" as the distinction.

There are a lot of other characteristics that people sometimes use to identify the classes. One which is kind of amusing, and about as accurate as any, is based on the way two married couples group themselves when they go out together in a car. Lower-middle-class couples will sit husband-husband, wife-wife; upper-middle-class couples sit husband-wife, husband-wife; the upper class exchanges wives. Now you know where you stand.

The shift in values that I'm talking about is occurring throughout our society, but more in the upper than in the lower classes. It is a shift from the values listed in items 1 through 13 of Chart 9 to the values named in items 14 through 26. The main movement in America today is this movement of changing values. Values are things which one can't say are right or wrong.

CHART 9   Some important human needs and/or values
1. Respect and recognition (competitive and mutual).
2. Proper mix of change, stability and/or continuity.
3. Rectitude, duty, responsibility (fulfilling ethical, moral and/or religious imperatives).
4. Daily activities and disciplines which are end as well as means — judged to be, in themselves, fulfilling and meaningful.
5. Having status — a recognized position, role, identity.
6. Advancement orientation — enhancing one's status.
7. Achievement (gaining and using skills, meeting challenges, solving problems, creating and/or doing worthwhile or admirable things.
8. Wealth (access to available resources).
9. Physical well-being (safety, health, comfort).
10. Physical power (over things — territoriality?).
11. Egoistic immortality (recognition).
12. Loyalty to or submergence in familial (shared fate, common commitment, ego-identification) structures.
13. Political representation (voting on and protection from community decisions).
14. Political power (over people and community decisions).
15. Praise, reassurance, attention, etc.
16. Justice to be done and/or morality to be made manifest — e.g., appropriate rewards and punishments for "good" and "bad" behavior.
17. Assurance and confidence about the important values.
18. Sensual satisfaction (food, sex, music, art, aesthetic and pleasant surroundings and experiences).
19. Adventure, excitement, danger.
20. Friendship, companionship, affection, love (to give and/or receive).
21. Enlightenment and understanding.
22. Play, spontaneity and self-expression (being oneself).
23. Having and sharing spiritual, mystical, religious experiences, codes and/or fulfillment.
24. Satisfaction of anger, revenge, other hostile emotions—perhaps slightly sublimated or masked.
25. Masochistic, sadistic, nihilistic etc., motivations — perhaps somewhat sublimated or masked.
26. Other "perversions" (sexual, gustatory, drug etc.).

A Japanese Society?

Let me digress for just a few minutes to the Japanese issue; then we'll come back to values.

Chart 10 gives twelve reasons why the Japanese economy ought to continue doing better than, say, European economies. Each of these factors helps to sustain a high growth rate—call it 8 or 9 percent better than the European rate. In the United States the average consumer saves about 7 percent of his income; the average Japanese consumer, about 17 percent. Out of the total gross national product Americans save maybe 15 or 16 percent; the Japanese about 35.

Now, just imagine that this country had an extra 22 percent of gross national product. That would be $220 billion which would go for investment right now. Just think what it would do to the country.

The Japanese have managed to have American-style mass education and European-style standards of quality through high school, and they are moving this achievement into the college levels. They are adequately capitalized. Their debt-equity ratios tend to be five to one, six

---

CHART 10   Twelve reason for probable continued growth of Japanese economy

1. High saving and investment rates.
2. Superior education and training.
3. "Adequate capitalization," in their terms.
4. Risk capital readily available.
5. Technological capabilities competitive with West.
6. Economically and patriotically advancement-, achievement-, work-oriented employees — loyal, enthusiastic and accustomed to deferred gratification. Trend will probably strengthen.
7. High morale and commitment to economic growth and to surpassing the West—by government, management, labor and the general public.
8. Willingness to make necessary adjustments and/or sacrifices.
9. Excellent management of the economy—by government, business and, to some degree, labor. Result is a controlled and, to some degree, collectivist ("Japan, Inc.") capitalism. But economy is still competitive and market-oriented (though not market-dominated).
10. Adequate access — on good and perhaps improving terms — to most world resources and markets.
11. Almost all future technological and economic — and most cultural and political — developments seem favorable to continuation of these trends.
12. Relatively few and/or weak pressures to divert major resources to "low economic productivity" uses.

to one, seven to one. By American economic standards, most Japanese companies would be insolvent, but they operate just fine. And this means that they can obtain their capital very cheaply, whereas an American company has to get at least half its capital in equity capital, which is very hard to raise. Japanese firms couldn't care less; risk capital is readily available.

Here I want to tell a story which is a little misleading, because it carries an implication about Europeans and about Americans which is not true. But maybe it is true about the kinds of Europeans and Americans who attend conferences in Japan—you know, top-notch, distinguished businessmen.

At a weekend conference in Japan about three years ago, I asked the Japanese who were present, "What does your family feel about your working weekends?" They replied (1) that they must be doing something very important, otherwise why would they be working weekends, and (2) that their family had to make it up to them because they sacrificed their weekend for the common good. I asked the Americans how, if we had been in America, would their families feel about their working weekends, and they said, "Well, we've got to make it up to the family, you know, for killing their weekend." And I asked the Europeans, how it would be in Europe, and they said: "The issue wouldn't arise."

Then I asked, "What would happen if an opportunity came up for your firm to invest, say, 5 percent of capital—a big hunk of its money—in a business opportunity, and it was your calculation that there was a 50 percent chance of tripling or quadrupling the investment in two or three years, and a 50 percent chance of losing everything. What would your firm do?"

The Japanese said, "We wouldn't believe the calculation." I said, "No, no, suppose you checked and rechecked and argued, and you do believe it." Well, they didn't understand the question, because as far as they were concerned, to believe the calculation was to make the investment. There could be no separation between the two concepts.

One of the Americans asked, "Would the loss be noticeable?" I said, "Very!" He said, "How noticeable would the gain be?" I said, "People would approve of it, but they wouldn't slap you on the back." He said, "We don't go." The name of the game for them is not to look bad, you see.

The European answer was: "The issue wouldn't arise."

Now, that overstates the European and American traditions. There are risk-taking European companies, lots of them; but not as many as you would have thought.

The Japanese got very upset with the American answer. They said, "But you Americans are great risk-takers." I said, "Their fathers were." And southwest Texas millionaires, mid-western American businessmen and occasionally others may be risk-takers today. But relatively speaking, the big, established American firms are not. There are exceptions; risk-taking is one of the main roles of the conglomerate, for example. And I'm sure AT&T has big capital investments which are occasionally risky. IBM, I think, put something like 3½ billion into the 360, which was a risk, in a sense. But largely speaking, the big American firms are slow to take big risks now.

Two more points on this issue, and then we'll leave the Japanese to their success.

First of all, the engine of growth in Japan is not profits; it's growth rates, market shares and prestige, and it also has to do with GNP per capita.

I gave a talk once to a very senior group in Japan—their equivalent to our National Security Council (they don't have one, but there's a committee of the Liberal Democratic Party). It was four, five years ago, before they had caught on to the idea that they could do very well. I asked, "Would you be doubling the size of your economy every six, seven years, if there were no West to catch up with?" They said, "Of course not. I mean, what are we, crazy?"

Then I asked, "All right, if the purpose is to catch up with the West, what's going to happen when you catch up with us?" They said, "That isn't going to happen, is it?" I said, "About the year 2000, is our guess." They said "No." I said "Yes." Calculations here are very uncertain, of course, but give or take ten years, you can make a persuasive case that Japan will pass us in this period.

CHART 11   Matsushita workers' song

For the building of a new Japan
Let's put our strength and mind together,
Doing our best to promote production,
Sending our goods to the people of the world
Endlessly and continuously,
Like water gushing from a fountain,
Grow, industry, grow, grow, grow!
Harmony and sincerity!
Matsushita Electric!

At the time, I was talking about per capita income. The current estimate is that they will pass us in per capita income in 1990 and in total gross national product in the year 2000.

They said "No." I said "Yes." They said, "What should we do?" I said, "You are the Japanese; you tell me." They said, "You've been thinking about it; you tell us." I said, "You might go back to Tokugawa Japan—the sword ceremony, tea ceremony." (This is what in a few moments I will call the tradition of the gentleman.) They said, "That's for tourists." I said, "You can't try conquest any more. That doesn't work." So there was a kind of a murmur as if to say: We'll re-examine that subject. As I said, very serious people.

Chart 11 gives a typical song that's sung by workers in Japan. Top management and labor get together and sing this kind of song for five or ten minutes every morning, and if you ask them why—they say, "We like it; it's fun."

It's dangerous to show this song to Americans; they look at it and want to throw up. IBM used to have a song very much like it, but they burned it in the late forties. You'll find some American sales organizations singing similar songs, off-key, in very thin voices. But there's nowhere in the world except Japan that this kind of song is sung enthusiastically. You may point out that in Vietnam when the VC take over a village, the villagers have to sing such songs for about an hour a day. Then the South Vietnamese take it over and the villagers sing for about a half-hour a day. But if you ask the villagers, "What do you want to do?" they'll say, "We want to quit singing." They don't do it for fun.

Now, what's interesting about this song of the Matsushita workers is that there's nothing wrong with it. It is not a gung-ho song. It is not a chauvinistic or parochial song. It is a simple set of declaratory sentences which are meticulously accurate. Take a look at it: Building a new Japan—every five, six years they double the size of the

---

CHART 12   A crucial Japanese concept

"There is at any given moment a definable world-ranking order of such character that as between any two nations one is always higher and the other lower. It is never the case that two nations stand on exactly the same level. Even when they appear close to each other, there is always a set of clues that allow the sensitive observer to discriminate between them and see their place in the ultimate ranking system."

— Herbert Passin

economy. Put their strength and mind together—they're unbelievably harmonious in that respect. Doing their best to promote production—that's correct, and it is product, by the way, not profits. Sending their goods to the people of the world—this is a slight exaggeration: 19 percent of their goods are consumed internally, and they're not as dependent as people think on foreign trade, but they do export a lot. Grow industry, grow. The company which sings this song, Matsushita Electric, has been growing 30 to 40 percent a year every year for the last twenty years. Harmony and sincerity; yes.

What's wrong with the song? Nothing. Why can't we sing it in the West? Low morale, that's all. It's not that the song is off; it's morale that's off.

CHANGING VALUES

To go back to the issue of changing values, Chart 13 lists the twelve values that I will call traditional levers in America: the sources, if you will, of reality testing, of social integration, of meaning and purpose. All twelve of these levers are disappearing in upper-middle-class America, or have already disappeared—particularly, say, among the kind of kids who go to a "prestige" school. If you ask a Harvard

---

CHART 13 The twelve traditional societal "levers" (traditional sources of "reality testing," social integration and/or meaning and purpose)

1. Earning a living.
2. Defense of frontiers (territoriality).
3. Defense of vital strategic and economic interests (or vital political, moral and morale interests).
4. Religion.
5. Tradition.
6. Other "irrational" and/or restricting taboos, rituals, totems, myths, customs, charismas.
7. Biology and physics (other pressures and stresses of the physical environment, the more tragic aspects of the human condition etc.).
8. "Martial" virtues (duty, patriotism, honor, heroism, glory, courage etc.).
9. Manly emphasis: In adolescence, team sports, heroic figures, aggressive and competitive activities, rebellion against "female roles." In adulthood, playing an adult male role.
10. "Puritan" ethic (deferred gratification, work orientation, advancement orientation, sublimation of sexual desires etc.)
11. A high degree (perhaps almost total) of loyalty, commitment and/or identification with nation, state, city, clan, village, extended family, secret society and/or other large grouping.
12. Sublimation and/or repression of sexual and aggressive instincts.

---

student today, there is no item on the chart, with the possible exception of "achievement orientation," which is important to him.

Contrast this situation, say, with that of Israel. The Israelis have a funny country: 40 percent European Jewish, 60 percent Oriental Jewish. The Europeans run the country; they're literate, white, tall. Oriental Jews are darker, illiterate, and short. The Israelis have all the problems we have with black-white issues. There's a lot of resistance building up among the Oriental Jews. Nevertheless, European Jews run the country, and they have insisted on a forceful Europeanization —very rapid.

Every now and then a Moroccan Jew will walk up to a European Jew and say: "I don't like your middle-class European bourgeois values. Leave me alone. What you're doing is 'cultural aggression'."

The European always answers: "We're surrounded by a hundred million Arabs."

The Moroccan Jew thinks about ten seconds and says: "Where's the engineering school?"

What can he do! You notice how it straightens him out.

That's what I mean by a lever. In this case it's reality testing, in two separate senses. First, you tell the Moroccan Jew that whatever his value system is, it better include European engineering, or he's dead. Second, he does not really prefer the Moroccan system to the European. What he was expressing was not deeply held values, but irritation, annoyance, frustration, a current fashion.

Now, I'm not saying that European values are necessarily superior. One thing which comes out very clearly from our studies is that worldwide, both inside and outside the West, there's going to be an increasing rejection of European value systems. But this doesn't hold for the Moroccan Jew. Like our own Black Power advocates, most of his rejection of the white world does not express internalized values; it expresses current fashion, if you will.

Take "earning a living," item 1 in Chart 13. You know what it costs to live in the United States if you're an upper-middle-class kid and want to live very cheaply? To live as a hippie, that is? Give a guess!

Until the last couple of years 95 to 98 percent of all the hippies were upper-middle-class kids. So there are few lower-middle-class kids among them except for those with emotional problems—you know, emotionally sick.

The standard figure for hippie living is ten bucks a week. Now, this is a figure which may not be right today, because hippies have tended to leave the cities, but it was right, say, in 1968 for Haight-Ashbury, for the East Village, for Cambridge, for Los Angeles. That's $500 a year, and it includes drugs. No money from home. So part of the name of the game is to play it poor.

It happens that the lowest salary Uncle Sam pays in the Post Office for the graveyard shift — 6½ hours' work a day, 32½ hours' work a week — is $500 a month. Hippie upper-middle-class kids, college-educated, as a rule have no trouble passing the exams. Whereas the typical applicant competing in the Post Office exam gets 60 or 70 on the tests, a hippie will get 95, 98, 99. It means that he can go to work in the Post Office any time he wants. He can then, living twelve to a pad, work one month on, eleven months off and live quite adequately.

Even today if you go to the Post Office in the four cities I've mentioned, about one-third of the people on the graveyard shift in jobs requiring literacy are hippies, complete with beads, sandals, no baths. A lot of them are doing it one month on, eleven months off, and to a man they resent the one month on. And of course, if you're eleven months off, that one month on is an intolerable chore. You're not used to it. It's a punishment. And the hippies are all in favor of a guaranteed annual income.

Another third of the workers on the graveyard shift are Negro. You might ask: What's a bright Negro doing on the graveyard shift? And the answer is: He has another job during the day, or he's going to school during the day; he's an upward-mobile young Negro. And the two groups hate each other's guts. There are no communication problems; they communicate just fine. The better the communication, the more intense the hatred. One group is throwing away what the other group is killing itself to get. Nothing racial; purely cultural; but they dislike each other with an intensity which would surprise you. They represent threats to each other, you see.

A minor point: About three years ago, the San Francisco Post Office put out a memo: "If you deliver the mail or otherwise meet the general public, you must wear shoes." Hippies protested that this was discriminatory, aimed directly at them—which was true. So the Post Office investigated and reissued the memo: "If you deliver the mail or otherwise meet the general public, you must wear shoes. This is not a change in policy; we would have mentioned the matter earlier, we just didn't understand that it would come up." Gives you a sense of change.

New Levers

All right, for whatever it's worth, none of the traditional social levers operate any more in the same way in upper-middle-class America. What is operating? The new levers coming up are shown in Chart 14. They're not bad, by the way. Take neo-Cynicism—I mean the philosophy of third-century Greece, brought up to date. You remember Diogenes, the man with the lamp looking for an honest man? He was a cynic. His father was a counterfeiter, and one way to describe Diogenes' entire life's work is to say he tried to prove that the world was more counterfeit than his father. "Cynic" means dog, and he lived like a dog: his private functions were performed in public; he slept in a jar, which was a Greek equivalent of a coffin. To many of us today he would be the spittin' image of a hippie or a new leftist, except that he was disciplined, aescetic and logical, which makes a difference. But otherwise he was a hippie. He was without question one of the great men of Greek culture.

The Greeks tell a story which illustrates their feeling about Diogenes. Alexander the Great, the master of the world, was supposed to have come to Athens specifically for the purpose of seeing the Greek

---

CHART 14 The current "transition" and/or search for meaning and purpose seems likely to encourage:

1. High consumption, materialism—mostly by bourgeois, relatively traditionalist middle class.
2. Neo-Cynicism.
3. Being "human" (neo-Epicureanism, familial and altruistic motivations and/or emphasis on interpersonal interactions).
4. Fulfilling a sense of responsibility (neo-Stoicism).
5. Neo-gentlemen (e.g., neo-Athenians and/or Europeanization of U. S.).
6. Self-actualization.
7. Special projects or programs that create general or specific esprit, élan, pride, excitement, charisma and/or chauvinism.
8. Humanist left, responsible center, conservationists.
9. "Bread and circuses" (e.g., both welfare and "happenings").
10. Rise of new and old cults.
11. Protest, revolution, violence as a kick or even a way of life — (e.g., commitment to nihilism, anarchism and/or neo-Fascism as well as "ordinary" protest movements resulting in continuing and continued demonstrations and riots).
12. Semi-permanent adolescence.
13. Other kinds of "dropouts" and quasi dropouts.
14. Emotional and "reactionary" backlash.

Cynic. They met in the square, and Alexander was very polite. He said, "You know, if I were not Alexander, I would be Diogenes. That would be my aim in life."

Diogenes was very polite—he said, "If I were not Diogenes, I would be Alexander."

Then Alexander asked him, "Is there anything I can do for you?" Now, this is the man who had conquered everything, but had been incredibly generous, had given everything away except power—you know, diamonds, jewelry, castles, slaves, wealth. When this man asked, "Is there anything I can do for you?" it was a serious offer.

Diogenes thought for a few seconds and said, "You're in my sun. Would you move over about two feet?" Do you get the quality of the man?—very impressive.

Being a human being—this is a big issue in America today. To be a neo-Epicurean—again, it comes out of the Cynic school, with no connotation here of the gourmet, but of a hedonist, hedonism in moderation. This means that the way you get satisfaction is by having very few demands. The Southern California barbecue culture is a modern example—though again, without the Greeks' discipline, aesceticism and logic. But the neo-hedonist, like the ancient hedonist, has no interest in advancement orientation, in public life, in fame, in achievement; he wants home, garden, porch, friends, family. A good deal of sensitivity training is based upon a belief that you have to start going in this direction. T-groups may or may not increase productivity, but they may make the office a better place to live in, if you like that kind of thing.

Neo-Stoics were a very important group in the Greek Empire. They also ran the Roman Empire for three hundred years. The tradition was handed down within families: "Other families may act that way; ours doesn't!" They were pessimistic people—pacifists who ran an army, cosmopolitans working for a nationalist empire, democrats working for an authoritarian state. Their attitude was not that of the soldier doing his duty, trying to win, but that of an actor in a play. It's not up to the actor whether he wins or loses; that's the author's privilege. All he can do is perform the part as well as possible. You can't be more pessimistic than that.

We generally think of the Roman Empire in terms of the decline of Rome. But when Nero was burning Christians in Rome, the outer reaches of Empire were being very well run by Stoics.

America has lots of Stoics. I sometimes give a talk like this one to colonels, and they all immediately identify as Stoics. They say, "You mean to say we've been doing our duty and getting ourselves killed and nobody's going to appreciate it?" I say, "That's roughly right." "You mean to say our kids are going to do the same thing and people are going to laugh at them?" I say, "You put it very well." Some of them say, "Maybe we won't do it." I always answer, "To use the current jargon, you're hung up." It's a depressing thought.

Actually, America produces Stoics like mad. For better or for worse, we can find people to do the unpleasant, hard, difficult, dangerous jobs, and do them for a long time to come. Now, it may not last indefinitely. When Marcus Aurelius died, he was the last Stoic with influence in the Roman Empire. From that point on, Rome looks like Saigon. But a three-hundred-year run is a long time, you know; not a short time.

Still another new social role will be that of the neo-gentleman. People talk about the Americanization of Europe, but there's a much more important process going on: the Europeanization of America. You find this particularly at the prestige schools. If you go to Harvard, don't look at the 5 or 10 percent who are dropout-leftist-protest people; look at the 80 percent who know how to ski, skin dive at three hundred feet, and write poetry. These people possess an unbelievable knowledge of both rock-and-roll and classical music, and they may also collect Chinese porcelains. They are "gentlemen," in a sense.

Chart 15 describes the same process a little more dramatically. It lists five ideologies which are characteristic of different phases of Western culture. When I was young, I thought my grandfather was down in the bottom part of column 5—pathologically submissive to God's will. In the last forty years I've decided that he was a very reasonable man.

We would argue that the Nazis were raised in no. 3, a Conscience ideology, at its pathological level of manifestation, and that most Americans in this room were raised with the values of Conscience or Reason as an ideology. But the hippie would argue that most Americans in this room were raised in something like the same value system as the Nazis. You know, it's all in the point of view.

The Impulse values in column 1 are those of childhood—very attractive in a five-year-old. You may or may not like them in a thirty-year-old. The hippie consciously says he wants to be a child; he's trying to preserve those values.

Now, if you ask the hippies or the New Left or the SDS people, "What is this society you're trying to make?" they say they don't

CHART 15 Social and political ideologies may emphasize:

1. IMPULSE	2. REASON	3. CONSCIENCE	4. TRANSCENDENCE	5. GOD'S WILL

Leading to, at best, a reasonable emphasis on:

Freedom	Rationality	Loyalty	Spirituality	Revealed truth
Spontaneity	Moderation	Dedication	Perspective	Absolutism
Creativity	Thoughtfulness	Tradition	Pan-humanism	Salvation
Perceptiveness	Meliorism	Organization	Idealism	Righteousness
Participation	Flexibility	Order	Altruism	Eschatology
Sensory awareness	Calculation	Obedience	Mysticism	Worship
Self-actualization	Planning	Self-sacrifice	Detachment	Awe
Joy and love	Prudence	Justice	Reverence	Submission

But with a corresponding potential for pathological degree of:

Permissiveness	Abstraction	Authoritarianism	Fatalism	Bigotry
Impulsiveness	Theory	Rigidity	Passivity	Fanaticism
Anarchy	Rationalism	Righteousness	Mysticism	Righteousness
Lawlessness	Indecision	Despotism	Naivete	Dogmatism
Chaos	Dehumanization	Sado-masochism	Unworldliness	Hypocrisy
Nihilism	Scientism	Punitiveness	Superstition	Superstition

know. Most of them claim total ignorance of what the new society should be like. They have some ideas—vaguely anarchistic but not very serious. Their position is that they are not Jesus Christ; they're John the Baptist. There's no point in asking John the Baptist what the message is; he hasn't got it. He's an honest man who'll tell you, "I don't have the message, but there's one on the way, so get that wax out of your ears."

I try to point out to hippies that revolution is the one understood process in history, and that it involves chaos, ordeal and catharsis. But what the hippie will not notice is that this is not a pleasant process. During the ordeal and catharsis, people work out the most unbelievably ingenious tortures. What are they trying to do? They're trying to get your attention. And you know, they do. I would just as soon not go through the process if at all possible, and I suspect we don't have to.

TECHNOLOGICAL CRISIS

All right, this is the basic picture. Let me hit one more chart, Chart 16, which will try to give you a feeling for our 1985 technological crisis. You understand, what I'm trying to do here is not deliver any specific message of my own but give you a gestalt view of what's going on.

Chart 16 mentions intrinsically dangerous technological possibilities and prospects. For example, in the 1980's atom bombs get incredibly cheap. To take another example, it's almost certain that within the next five to ten years we'll be able to choose the sex of children, although the methods will be fairly crude. It will be done either by artificial insemination or by checking the sex of the baby—and giving

---

CHART 16   1985 technological crisis areas

1. Intrinsically dangerous technology.
2. Gradual and/or national contamination or degradation of environment.
3. Spectacular and/or multinational contamination or degradation of environment.
4. Dangerous internal political issues.
5. Upsetting international consequences.
6. Dangerous personal choices.
7. Bizarre issues.

a pill in the first or second week of pregnancy to abort the unwanted child. As my European friends put it: You Americans may use it; it's much too gadgety for us. They show no interest in it.

By the end of the century, we may be able to do the same thing cheaply and simply, perhaps with a pill or a salve. Now, let's assume that scientists find such a pill. I would argue that they would be wrong to publish the discovery. I've never in my life been in favor of an index of forbidden knowledge before, but let me tell you the effect of this pill. In countries like India or China the sex ratio would go from 51:49 to 99:1. Indian and Chinese peasants have no interest in female children. Obviously that would help solve the population problem, but believe me, it would raise more problems than it solved. And by the way, peasants will not listen to advice from their government. I'm sure that after fifteen or twenty years, they'd catch on that they were oversupplied with males and short on females, and make a wild swing the other way. But these things are not conducive to orderly development of a community.

I'd bet that places like, say, Japan and West Germany would go to an 80:20 ratio in favor of males. In the United States I've checked with a lot of people, taken an informal poll, and Americans tend to like the pattern of boy, girl, boy, which is 2:1.

What I'm saying is that once this kind of pill is widely available, the government is almost surely going to have to interfere in a very personal matter. It raises peculiar issues.

Take genetic engineering. Today we do a certain amount of it. That is, some doctors will take the amniotic fluid of a pregnant woman to discover if there are certain kinds of defects they can fix before the baby is born. This is a useful technique.

Very soon most of my friends in this field believe that we'll be able—within the limits of the germ plasma—to pick the height of the baby, its vitality, its IQ, its skin color, hair color, eye color—you know, general kinds of things. Anybody think the world will be happier? I don't.

There was an experiment several years ago which I found very interesting. It went under the name of "vegetative reproduction." The scientist took chromosome material from the intestines of a frog and put it in a fertilized egg of a different species of frog. He reproduced the genetic twin of the original frog. Now, if this works out, and you can do it for any kind of mammal, within a decade or two you may be in a position to produce the genetic twin of any individual. My wife keeps a technology file for me, and when I mentioned this possibility to her she got very upset. I was kind of pleased about the idea. We both had exactly the same picture: ten Herman Kahns a year. Her position is, she wouldn't live in that world; my position is, you don't want to run out. These are issues of taste, if you will.

My favorite example of technological possibilities is also the most dramatic. The first experiment was done about twelve years ago. A scientist took a "pleasure center" in the brain of a rat, wired it to a lever, which is a simple operation, and gave the rat the choice of pressing the lever or of having food, water, sex or rest. Some 6,000 times an hour the average rat pressed the lever and ignored his other choices. If you forced him to take a little food, a little water, a little rest—he doesn't need sex, he's got something better—he leads a longer, and as far as we can tell, happier and healthier life than the control rats.

And we have verification that it's really, truly happiness. If you make the rat cross an electric-shock region to measure the intensity of his drive, he will take unbelievably heavy shocks to get to this lever. Or if you teach him to go through a maze, and if you use rewards to reinforce his maze learning, this lever is the most effective reinforcing device we know of in teaching him. There have been a lot of experiments of this kind done on animals. Some that have also been tried on human schizophrenics and paranoids gave confusing results, as well as raising the question of whether you should do this to human beings.

But let's assume that a human being has ten pleasure centers. You get them wired to a console in your chest. It's a very simple operation and its revocable—it can be undone. Now, I don't think you should play your own console—I'm a very square guy. On the other hand, I don't want to be rigid about the matter. Let's say any two consenting adults can do it. Imagine the conversation: "You ever try 3 and A together?" "I want you to get set for an honest to goodness mind-blowing experience." "Brrrrrr!"

And by the way, if the animal experiments are any clue, this is not addicting; it's habituating. You know the distinction? Marijuana is habituating; cocaine is addicting. Let me ask the following: How many people here would try it just once?

This is the kind of thing which is beginning to turn the younger generation off, not on. They don't like this world. For the first time in American history—in fact, in Western culture—you've got a large group of young people who are not arty, not aesthetes, yet are anti-progress; and their argument is not completely without merit.

HERMAN KAHN

Herman Kahn is well known for his predictions of the political, economic, technological and cultural changes the world will face in the next ten to thirty-five years. He is recognized as one of the most brilliant military strategists in the United States. As director of the Hudson Institute, he is especially concerned with problems of national security.

By training Mr. Kahn is a physicist. He studied at the University of Southern California and then, after two years in the army during World War II, completed his undergraduate studies at UCLA in 1945. He received a master's degree in 1948 from the California Institute of Technology.

That year he was appointed senior physicist and military analyst at the RAND Corporation. One function of his job was to conduct lectures for military and aviation leaders, including some of America's chief policy makers. In the fifties his principal area of concentration was applied mathematics, and during this period he became an expert on weapons design. Publication of his *On Thermonuclear War* in 1961 established him as a national figure. In 1961 he left RAND to become co-founder of the Hudson Institute.

During his career Mr. Kahn has been involved in studies on Latin America and alternate world futures, and has done research in strategic warfare and basic national security policies. Most recently, he has been involved in studies of the corporate and government environments in the period 1975-1985. He has served as consultant for various government and private organizations, including the Gaither Committee on Civil Defense and Strategic Warfare, the U.S.A.F. Scientific Advisory Board, the Atomic Energy Commission, Oak Ridge National Laboratory, the Office of the Secretary of Defense, the State Department, and the Stanford Research Institute. Currently he serves on the New York State Study Commission for New York City.

Herman Kahn has written five books and numerous articles for scientific and general magazines, including *Fortune, The New York Times Magazine, The Saturday Evening Post, Bulletin of the Atomic Scientists, Daedalus* and *Commentary*. As a lecturer he has spoken at most U. S. Service colleges, at various American universities and at universities and research and defense centers around the world. He is also a member of the Council on Foreign Relations, the Center for Inter-American Relations, and various professional honorary organizations.

BIBLIOGRAPHY

On Thermonuclear War. Princeton, N. J.: Princeton University Press, 1961.

Thinking about the Unthinkable. New York: Horizon Press, Inc., 1962.

On Escalation. New York: Frederick A. Praeger, Inc., 1965.

and A. J. Weiner. *The Year 2000.* New York: Macmillan Company, 1967.

The Emerging Japanese Superstate. Englewood Cliffs, N. J.: Prentice-Hall, Inc., 1970.

SUPPLEMENTARY CHARTS

The following charts offer something more to think about. Note that charts throughout this presentation are the responsibility of the author; no conclusion should be attributed to the Hudson Institute or any associated agencies. Items may be unintelligible or directly misleading if they are interpreted outside the context of the talk they accompany.

CHART 17   Three standard cultural phases

SOROKIN	SPENGLER, TOYNBEE, et al.
1. Ideational	1. Growth, spring, childhood
2. Idealistic	2. Maturity, summer
3. Sensate	3. Autumn, winter, decline

SCHUBART	BERDYAEV
1. Ascetic-messianic	1. Barbaric-religious
2. Harmonious	2. Medieval-renaissance
3. Heroic-Promethean	3. Humanistic-secular

There are, of course, always minor themes, reluctant or nominal conformers, undergrounds, holdovers, dropouts, dissenters, schismatics, heretics, unbelievers and other deviants or exceptions. (These may be a majority of the population, but still not very visible.)

CHART 18   One could contrast the ideational, integrated (idealistic) and sensate systems of:

Fine arts	Family relationships
Truth systems	Civic relationships
Music	Literature
Performing arts	Ethics
Architecture	Education
Law	Government
Economics	Etc.

One must, of course, also worry about the theory of consistency and leading and lagging sectors, as well as the exceptions previously mentioned.

CHART 19 Four cultural stages in the fine arts

IDEATIONAL ART	IDEALISTIC OR INTEGRATED ART	SENSATE ART	LATE SENSATE ART
Transcendental	Mixed style	Worldly	Underworldly
Supersensory	Heroic	Naturalistic	Protest
Religious	Noble	Realistic	Revolutionist
Symbolic	Uplifting	Visual	Overripe
Allegoric	Sublime	Illusionistic	Extreme
Static	Patriotic	Everyday	Sensation-seeking
Worshipful	Moralistic	Amusing	Titillating
Anonymous	Beautiful	Interesting	Depraved
Traditional	Flattering	Erotic	Faddish
Immanent	Educational	Satirical	Violently novel
		Novel	Exhibitionistic
		Eclectic	Debased
		Syncretic	Vulgar
		Fashionable	Ugly
		Superb technique	Debunking
		Impressionistic	Nihilistic
		Materialistic	Pornographic
		Commercial	Sarcastic
		Professional	Sadistic

CHART 20 Three systems of truth

IDEATIONAL	SENSATE	LATE SENSATE
Revealed	Empirical	Cynical
Charismatic	Pragmatic	Disillusioned
Certain	Operational	Nihilistic
Dogmatic	Practical	Orwellian
Mystic	Worldly	Blasé
Intuitive	Scientific	Transient
Infallible	Skeptical	Superficial
Religious	Tentative	Weary
Supersensory	Fallible	Sophistic
Unworldly	Sensory	Formalistic
Salvational	Materialistic	Atheistic
Spiritual	Mechanistic	Trivial
Absolute	Relativistic	Changeable
Supernatural	Agnostic	Meaningless
Moral	Instrumental	Alienated
Emotional	Empirically or	Convenient
Mythic	logically verifiable	Absolutely relativistic

CHART 21 Six economic groupings in year 2000 (millions of people)

VISIBLY POST-INDUSTRIAL
($7,000/capita & over)

Japan	125
U. S.	305
Sweden, Denmark, Norway	20
W. Germany & W. Berlin, France	130
E. Germany & E. Berlin	20
Canada	30
	630

EARLY POST-INDUSTRIAL
($4,000-6,999/capita)

United Kingdom	60
Benelux, Australia, Switzerland, Finland, Europe N.E.C.	50
Czechoslovakia, Hungary, Bulgaria	40
Israel & Australia	25
Italy & Yugoslavia	85
Puerto Rico, N. America N.E.C.	5
U.S.S.R.	315
	580

MASS CONSUMPTION

New Zealand	5
Poland, Romania, Albania	75
Spain, Greece, Portugal, Ireland, Iceland	60
Venezuela & Neths. Antilles	35
Libya, Jordan, Cyprus	10
S. & S.W. Africa, Iran, Taiwan	115
	300

MATURE INDUSTRIAL

Mexico, non-Latin Caribbean N.E.C., Brazil, Argentina, Colombia, Guianas, Surinam	455
Turkey, Iraq, S. W. Asia N.E.C.	140
Thailand, Malaysia, Middle S. Asia & E. Asia N.E.C.	190
Zambia	10
	795

TRANSITIONAL

6 C. American reps., British Honduras, Cuba, Dominican Rep., S. America N.E.C.	150
U.A.R., Morocco, Tunisia, Cape Verde I., N. Africa N.E.C., Syria	175
Mainland China	1,315
India, Pakistan, Ceylon	1,305
S. Korea, S.E. Asia N.E.C.	510
Rhodesia, Black Africa N.E.C., Oceania N.E.C.	265
	3,720

PRE-INDUSTRIAL

Sudan, Nigeria, Ethiopia, Tanzania, Uganda, Malawi	290
Burma	55
	345
World	6,370

(N.E.C. – Not elsewhere counted)

CHART 22   Traditional values fulfilled by work

BASIC ATTITUDE TOWARD WORK	BASIC ADDITIONAL VALUE FULFILLED BY WORK
An interruption	Short-run income.
A job	Long-term income – some work-oriented values (one works to live).
An occupation	Exercise and mastery of gratifying skills – some satisfaction of achievement-oriented values.
A career	Participating in an important activity or program, much satisfaction of work-oriented, achievement-oriented, advancement-oriented values.
A vocation (calling)	Self-identification and self-fulfillment.
A mission	Near-fanatic or single-minded focus on achievement or advancement (one lives to work).

CHART 23   Some reasons why Americans might reject current work-oriented, achievement-oriented, advancement-oriented attitudes

1. Why not?
2. Besides, anyone can make $10–$25,000/year by coasting.
3. There will be a minimum income guaranteed by the government – as well as other free and welfare benefits.
4. It will be easy to obtain an additional $1–$10,000/year from relatives or friends.
5. The marginal utility of money will go down.
6. Society will feel it can afford slackness and deviation.
7. Effects of changed child-rearing patterns.
8. Excessive reactions – intellectuals and beatniks – against "bourgeois," "managerial," "bureaucratic," "industrial," "puritanical," and "pre-affluent" values – the idea of the sheer indecency of getting richer – of miserly behavior.

CHART 24 Recent ideological trends

1. **1916–1940:** World War I provided vindication of pacifist and relativist critiques. Depression supported reformist critique and gave power to intellectuals – weakened "establishment" authority.
The intellectual anti-war literature of the period powerfully reinforced such attitudes as relativism, cosmopolitanism, pacifism, anti-militarism, reformism.

2. **1941–mid-1950's:** Hitler, Stalin, World War II, Cold War, and Korea undermined relativism, cosmopolitanism, pacifism, anti-militarism, and violent reformism. Nevertheless, 1948 Wallace movement provided example of American intellectual reassertion of those views.
1950's intellectuals withdrew. Students (i.e., "silent generation") responded to attitude of their teachers.

3. **Late 1960's:** World War II experience and lesson had receded into past – there was revitalized relativism – our system was no longer seen as good for its own sake.
Nuclear weapons and Vietnam War revitalized pacifism, anti-militarism and cosmopolitanism.
Civil rights and poverty causes revitalized reformism.
Inevitable political defeats of intellectual causes revitalized elitism and violence.

4. **1970's:** Promise to be "worse" than 1916–1940.

CHART 25 New elements in the politics of the 1970's

Humanist left: Interested in human relationships—i.e., man to man, man to woman, problems of alienation etc. Violently anti-hierarchy and anti-bureaucracy, thus favors participatory democracy and often tends toward anarchy. This group contains most current dropouts (hippies, new left etc.) and revolutionary student movements, but is much broader and has many fellow travelers. It has also inherited many former liberals.

Responsible center: Like the humanist left, a progressive and reformist group, but one which ranges from pure technocrats concerned only with keeping the system going (and who are skeptical of humanist-left capability to use power wisely) to those who truly value "the system." Relatively close to today's liberal consensus, but greatly modified by an understanding of the arguments for conservatism. While often emphasizing that they share many values and other characteristics of the humanist left, the responsible center tends to define itself as an opposition or alternative to the humanist left.

Alienated (reactionary?) "lower middle class": This is middle America or the "silent majority." Extremely conscious of being alienated from the liberal consensus and the "progressive establishment"; upward-mobile but financially hard-pressed; relatively racist but unwilling to use racist arguments. Enjoys hunting, non-intellectual, non-sophisticated TV and movies; democratic ways; manly behavior; patriotic and religious values.

Conservationists: A group which is literally conservationist in the sense of trying to conserve old values, but is not necessarily interested in the economics and political emphasis of the "Landon" or "Goldwater" conservative. Highly skeptical of technological, cultural and political progress, these new conservatives emphasize many of the same "old-fashioned" values and ideologies as the "alienated lower middle class" without necessarily having much tendency toward being racist, reactionary, authoritarian, chauvinistic or anti-intellectual.

CHART 26   Changing politics—a new political milieu may characterize the decade

1. A "thousand-year" secular trend in Western culture (see Chap. 1 of *The Year 2000*). Toward a sensate culture emphasizing cosmopolitan, humanistic, anti-militaristic, nationalistic, intellectual, relativistic, scientific, rationalistic, manipulative, secular, and hedonistic values, and soon the onset of the post-industrial culture.
2. A revival in the West of the post-World War I reaction against nationalism and militarism.
3. A current reaction (as exemplified by the New Left, but much more general) against modern science and technology, economic and administrative efficiency and private and governmental bureaucracies.
4. A crisis of liberalism characterized, among other things, by a reaction against individualism and rationalism and a rebellion against domination by much the same groups.
5. Increasing role of the intellectual, with a concomitant tendency toward removal of all "irrational" and restricting taboos, totems, myths and charismas; and toward a questioning of all traditional claims, facts, assumptions, and loyalties; an emphasis on the new and a rejection of the old merely because it is old, unless it passes. "Acceptable" intellectual criteria.
6. A reaction, both domestic and foreign, against the U. S. government – an inevitable result of the superpower status of the United States and the excessive postwar dependence on it.
7. General decline in reputation and prestige of armed forces and governing establishments – and U.S. Armed Forces and U. S. government establishment in particular – exacerbated by many seemingly incompetent (as opposed to "immoral") aspects of many recent incidents.
8. Continued, even stimulated, rising expectations (internally and externally) and a lower tolerance by intellectuals and upper classes generally of the existence of "irrational," "indefensible" and "unjust" inequities – complicated in many nations by upper- and upper-middle-class "guilt complexes" and anti-anti-left ideologies.
9. A surprisingly intense generation gap. In addition to the reasons above, affluent classes tend to raise their children in a permissive, gratification-oriented, passive, over-stimulated, "Now!" environment that contrasts markedly with the Puritan ethic (as well as with the depression and World War II milieu in which the current 40- to 50-year-old "governing class" was raised), while erosion of the twelve traditional levers is an enormous change in external environment.
10. In the United States a spotlight on – and start of – the resolution of such "societal failures" as Negro aspirations, persistent poverty, pollution, urban difficulties of various sorts, tending to cause an over-emphasis of these issues, unrealistic expectations, and a subsequent frustration, disillusionment and/or alienation.
11. Additional alienation of many upper and upper-middle-class youth – stimulated by seeming apathy and callousness of older generation toward unresolved issues of nos. 6-10 above, an anarchist-like reaction against bureaucracy and the system (which is often simply an extreme version of the first five trends) and a more or less normal cutting of adult apron strings – a cutting that, in the current milieu, seems likely to result in some relatively long-lived and rather eclectic sets of extremist reactions.
12. The various effects of current news media (particularly on the reporting of governmental violence – both internal and external). Although it does indeed tend to intensify and exacerbate almost every one of the above trends, the influence can be exaggerated. While it can make relatively minor issues and mistakes into disasters, in many cases it seems to be as much the ineptitude of the authorities as the nature of the reporting and of the medium that causes the problem.

9. CHOICES FOR TOMORROW

ARTHUR E. JENSEN
Professor of English
Dartmouth College

Even to speak of choices for tomorrow may seem to argue at best incurable optimism and at worst blind complacency. Today, for perhaps the first time in human history, we are asking ourselves: Will there be a tomorrow? And if there is, will it be shaped by the choices made by individual men? For some the outlook appears so bleak that they are retiring into a private world, living today as though there would never be a tomorrow, convinced that nothing can be done to divert man's headlong rush towards destruction, since the forces shaping the future appear unmanageable. But most of us are only too painfully aware of the need to choose. We feel a terrible sense of urgency and a terrible burden of responsibility. Will there be a tomorrow? Well, not unless we, individually and collectively, make the right choices—and make them now. Where are we going to find the wisdom to make those right choices before time runs out on us?

"We live in a revolutionary age." We have heard the statement so frequently that it has become a cliché, and our sensitivities have become dulled to its implications. Nevertheless, it is true. A great part of our problem is that we are living in a rapidly changing world but thinking, by and large, along traditional lines. As Robert Oppenheimer put it:

> This world of ours is a new world, in which the unity of knowledge, the nature of human communities, the order of society, the order of ideas, the very notions of society and culture have changed, and will not return to what they have been in the past.

And the very real dangers that confront man as he moves into this new society were suggested by Norbert Wiener when he wrote:

There is a real possibility that changes in our environment have exceeded our capacity to adapt. The real danger at the present time—the danger of thermonuclear war, the computing machine sort of danger, the population explosion danger, the danger from improvements in medicine . . . all these dangers make one wonder whether we have not changed the environment beyond our capacity to adjust to it, and whether we may not be biologically on the way out.

When the first atomic bomb was exploded in 1945, most men recognized, if only intellectually, that, in military matters at least, all humanity was entering a new age. To this day, however, most of us have failed to grasp emotionally the difference between a war, however terrible, out of which all but a small fraction of humanity will survive, and one in which there will be no survivors, or at best a miserable few waiting for their own deaths. We still somehow think of a nuclear war as a worse war, because our thinking still binds us to the past. In the past our conception of the future was always of one in which change was incorporated into a deeper changelessness. That is no longer true. If a human society exists in 2071, it will bear less relationship to that of 1971 than our society does to that of the Middle Ages.

Will there be a viable tomorrow? That the future is fraught with mortal dangers is increasingly apparent. There are tremendous forces working against the well-being and even the survival of man. One is overpopulation. "The earth has cancer, and the cancer is man." We are all familiar with those words of Aldous Huxley when he joined with other experts in predicting that, unless the population explosion is stopped and the trend reversed, there is no decent future for any life on earth. So, if nuclear war does not destroy us, we may well in another century breed ourselves into a world that is horrible to contemplate. If not, we may commit slow suicide by perverse use of the tools that technology has provided: nerve gas, the end of the road in antibiotics, or the irreparable pollution of our living space.

There are not only these cosmic dangers facing us; in the midst of the most affluent society in history, we are also witnessing a profound social and psychological discontent. This fear of the future comes at a period of greatest material wealth. Affluence has not brought social content. In fact, it seems that affluence may be our undoing and we may well choke on our own possessions. What is the use of an automobile if it must creep along our jammed highways, polluting the air we breathe, and if we can find no place to park it near our destination? What is the use of a beautiful apartment on Park Avenue if we cannot breathe the air that comes in the window? What is the use of providing a full table of food for every family if the food is contaminated by mercury or DDT? If the public sector of our society continues to deteriorate at its present rate, private affluence will soon cease to have much meaning. Affluence, it would seem, has brought social tension rather than social peace. As John J. Corson wrote in *Business and Humane Society:*

> The agonies of the late 1960s—threatening division among our people, riots in the ghettos and on the campus, the spread of drug usage, rising crime rates, and persisting and deepening racial tension—demonstrate that a society with generally comfortable or high personal incomes, generally profitable and expanding business enterprises, and a gross national product at peak levels is not the end-all of American objectives.

We have learned that we have been sailing in what we thought was a fair wind, but without clear values to steer by, and suddenly we find we are rapidly approaching catastrophe. One of President Nixon's speech writers recognized this prospect when he wrote for the State of the Union message to Congress in January 1970:

> In the next ten years we shall increase our wealth by 50 percent. The profound question is—does this mean we shall be 50 percent richer in a real sense, 50 percent better off, 50 percent happier? Or does it mean that in the year 1980 the President standing in this place will look back on a decade in which 70 percent of our people live in metropolitan areas choked by traffic, suffocated by smog, poisoned by water, deafened by noise, and terrorized by crime?

Is this what we have called progress? For two centuries we have equated progress with growth and have been dominated by a growth complex. This was a belief that economic growth was an unquestionable goal, both for the individual and for society. In almost every aspect of our lives we have looked for a rising curve: greater population; greater production of cars, shoes, TV sets; greater per capita income; greater spread of consumer goods. We wanted to grow. Growth came to assume almost a moral imperative. We now realize that to achieve this growth we have overexploited the resources of the planet, we have bred ourselves into numbers that threaten mass starvation and the elimination of many other forms of animal life on earth. We have polluted, in some places irreparably, the only home the human race has, this small spaceship of rapidly depleting finite resources. We have been acting as though we did in fact believe that this was the last generation. We used to think that pure air and water were inexhaustible. We now realize that we have in many ways upset the delicate balance of the biosphere and that, unless we make decisions now to change our ways, the future looks pretty desperate.

Obviously the first morality of any man who looks toward the future is to face these grim and desperate realities that confront us. For an hour this afternoon, however, I choose to ignore those possibilities, although many thoughtful men regard them as probabilities. Let us say that Armageddon can be avoided; that war, plague and famine, the old Malthusian triad, are held at bay; that a future is possible. What are some of the choices we have to make?

Before we can talk of choice we have to consider purpose, for the policies of any institution, the choices it makes, are determined by its overall purpose. What in this age do we find to be the overriding purpose of American society? Is it a little rough to say that for a generation it has been to check the spread of communism abroad and to increase private affluence at home? That is a far cry from what in the nineteenth century was unblushingly called the American Dream. A dream which, though we realize it was often distorted and even corrupted in practice, was still an ideal for the establishment of a just and egalitarian society on this almost virgin continent.

Somewhere, in our pursuit of affluence, that ideal lost its glamor. The post-war prosperity was not a time to bring it up. We were in the midst of an economic boom that tended to mute philosophic doubts. Few people in the 1950's were concerned about pollution or even overpopulation. Our national purpose during those years seems to have been only the containment of communism, with Senator Joe McCarthy leading the lunatic fringe of that policy. Our domestic policy seems to have been concentrated on not rocking the boat and on sustaining prosperity.

In what might be called an ineffective countercurrent of the 1950's, many intellectuals were deeply concerned by what seemed an absence of purpose in American life. They influenced President Eisenhower to appoint a commission, headed by Henry Wriston, to study and report on what in this generation should be our national purpose. They issued their report, a fat book called *Goals for Americans.* It was reviewed in the *New York Times* and a few periodicals. Then it quietly died. It was followed by an even fatter work sponsored by the Rockefeller Brothers Fund called *Prospects for America,* and that in turn was followed by a series of articles in *Life* magazine, later collected in a paperback, called *The National Purpose.* Both reports—and they were full of insight and understanding—were received with the same acute apathy that greeted the Wriston report. We were drugged by our own affluence.

Basically our society has confused ends and means. The means of a rich life became ends in themselves, and proliferated out of control and with little regard for ends. The example that comes most readily to mind is the explosive advance of science, and it is on the implications of that explosion that I want to spend much of my remaining time.

Up to now science has operated on what has been called the Everest Syndrome. Remember that on being asked why he wanted to climb Mount Everest, Mallory replied, "Because it is there." The question the scientist and technologist have asked is not whether something should be done but whether it can be done. The mountain is there to be climbed. René Dubois quotes "an eminent American scientist" as saying, "We *must* go to the moon for the simple reason that we *can* do it!" The acceptance of a challenge simply because it is a challenge is a noble characteristic of man, which in the past has led to great achievements and benefited the human race. We have maintained that assumption. Consequently, we have tended not to question the ultimate purposes or ends to be served by scientific research and technological progress. We have assumed not only that knowledge is good, but that real progress can be made only by increasing and exploiting knowledge.

We are no longer living with that assumption. Today we find among many men almost a revulsion against the scientific achievement that has made the modern world possible. They look upon the advancement of science as having, to an exponential degree, magnified the horrors of war, polluted our air and waters, and degraded man to be the mere tool of the machine. So into the worship of science has crept a rank heresy: It might be better for mankind if we lacked some of the knowledge that science has given us in the past and promises to give us in the future.

Take one or two examples. Medical research has developed drugs that will keep a terminally ill person alive for weeks and even months after he would normally have died—to his greater suffering and that of his family. Which opens up a basic question: Is it the doctor who has the right of decision in such a case? The patient? The relatives?

Do you know the word "cloning"? It is a process, mysterious to me, by which it is possible to create an exact duplicate of a living animal. Scientists have been able to do it with frogs, and some say it is only a matter of time before it can be done with human beings. But are we ready to make wise use of that kind of knowledge if it becomes available? Who decides which persons should be duplicated? Or do we just say that we *must* do it because we *can* do it? What about other probable scientific breakthroughs, where the implications are subtle and equally unpredictable? Some biologists say that within a few years, if we support the necessary research, it will be possible for a couple to decide in advance whether the child they are conceiving is to be a boy or a girl. Suppose that knowledge is achieved and all mankind has access to it. What are the predictable results? It is not altogether clear. The way nature does it is to toss a coin in each case, and the result is an almost perfectly even distribution of the sexes.

A sociologist took a poll and found that the average American family would want, if they had three children, two boys and a girl. If two children, one of each. If one, a boy. In some parts of the world the preference for male children is intense. In places like India the population of boys would become overwhelming. That topic has been mentioned before in this series, and I shall drop it except to say that we can all recognize that, if man were to impose his discretionary power instead of accepting the present natural distribution, there would inevitably be wild fluctuations in sex distribution over the generations, and consequent social dislocation. Just what would happen is not quite predictable, but we know that the widespread knowledge of sex determination would certainly not make for social stability.

Scientists may soon be able to determine the precise areas of the brain which control certain aspects of behavior, and by use of specific drugs they may be able to control that behavior. Shall we force such drugs on habitual criminals, on homosexuals, or on people who have unorthodox views? Suppose we had a drug—and it is here—that makes a person placidly adjusted to his environment. Undoubtedly it would be conducive to social peace, but at what price?

Sometimes we don't know the price of a heralded gain, even in the physical sciences, until years later. Twenty-five years ago we had a great scientific breakthrough with the discovery of DDT, the insecticide that was to perform miracles for farmers and housewives. Only now, a generation later, we find that DDT was not progress but a mistake that is exacting a terrible price from us and our descendants. The Pill was introduced to American women long before we had any knowledge of its possible side effects. We are beginning to question whether scientific progress inevitably serves the good of man.

The point is that scientists do not seem to have any peculiar wisdom about the long-range effects of their work. Nor about the ends and goals of society. Clemenceau said that war was too important to be left to generals. So it may be that the control and uses of scientific discovery are too important to be left to scientists. What I am suggesting is that the choice we have to make is whether or not there is to be a moral and political control over the direction of science and its exploitation. By moral I mean a clear discrimination on the basis of values between the triumph of attainment and the effect of that attainment on society. To go on assuming that we *should* do what we *can* do amounts to a moral abdication, a refusal to discriminate in values.

I don't think that issue has been placed before our society more clearly or concretely than in the controversy over the SST. We are familiar with the arguments. It can be done, so we should do it. If we don't, the Russians will. But for the first time society, through the American Congress, said that, on the basis of its overall effect on mankind, we had better stop this possible achievement in science and technology. Too many questions were unanswered. How destructive would the sonic boom be? How much damage would be done to the life-sustaining upper atmosphere? Would the price of getting a few wealthy people from New York to London an hour or two faster be just too high for society to pay? It was not a decision to spend or not to spend some money. It was a decision based on values. Whether the decision was right or wrong is beside the question. It is enough to say that for the first time a definite restraint was placed on a possible scientific advance, and it is a precedent that will, I am sure, be followed increasingly in the future. It was a dramatic vote, because both Congress and the people sensed the seminal importance of the decision.

Our whole tradition of the glorification of science and technology has made it understandably difficult to think of putting curbs on human exploration into new areas, even though we sometimes try to mitigate the results. We try to prohibit the use of DDT, but draw back from the logical step of prohibiting its manufacture. We publish warnings against the effects of smoking, yet I have heard no responsible voice suggest the next logical step, to prohibit the manufacture of cigarettes. In this area society has to break new ground.

But there are going to be some very tough questions and options. To forbid the manufacture of DDT or the completion of the SST is one thing. And easily effected. To tell the biologist in his laboratory that he must not indulge any more in research on the determination of sex is an intrusion into an area that has hitherto been held to be almost sacrosanct. It is to create rules and restrictions that are patently unenforceable. In the one case we act against a concrete object that may bring more harm than good. In the other we are attempting to restrict human thought, which we can't do; or to restrict the publication of that thought, which runs counter to all our traditions as a free people. Moreover, we know that our fabulous standard of living has been made possible only because we allowed scientists free rein. But the inherent dangers of the direction in which we are going make us pause.

It is easy to be simplistic in our thinking, as some have been, and to say that we must have a moratorium on science and technology because their development has been the source of so many of our woes. It is the swiftly moving industrial development, sprung from the loins of science, that has depleted our natural resources and polluted the planet. It is the application of the new scientific knowledge in biology and medicine that has brought on the population explosion. It is science that has made the old, stabilizing religious beliefs no longer tenable, and has thrown us into a society rapidly becoming secularized, with no supernatural sanctions for morality. It is science that has given evil men and nations the power to destroy the world.

Scientists and engineers have often led us astray. It is now becoming increasingly clear that the Aswan Dam, instead of being a blessing, is on its way to becoming for Egypt, and even for its neighbors, an unmitigated disaster. So might well be the projected sealevel canal across Panama. When we deliberately alter the balance of nature, disturb the great chain of being that has been established for millions of years, we have to make sure that the side effects of what we do will not be disastrous. With hindsight we can see that every so-called advance has had its price, and had we known how terrible the price would be we would have hesitated to use DDT, phosphates in detergents or lead in gasoline. Or, let us say, to construct the St. Lawrence Seaway.

Let us take a more complex issue. The population of South America is getting out of hand. It has been estimated that the GNP of South America is growing at the rate of 2 percent a year. The population is increasing at the rate of approximately 3.5 percent a year, which means that each year there is a 1.5 percent decrease in the standard of living among people who, for the most part, are barely subsisting. Why those wretched conditions are not even worse is accounted for in part by the high infant mortality rate in the continent. Suppose we found a way to cut in half the infant mortality of that region. We might quickly have a 4 percent rise in population and a 3 percent decrease in the standard of living each year. We know that much of the population explosion in other parts of the world is the result of introducing American medical and sanitary measures. The draining of swamps, the institution of insect control, and other such practices have resulted in a few generations in tripling the population on lands that could barely support their original numbers. Some of you have undoubtedly read the book *Famine* by the Paddock brothers, in which they indicate that in a few years we shall have to make a deliberate decision to let many of these areas starve to death.

Back to the South American babies. Shall we say that we have the power to save their lives but are deliberately letting them die? Tell that to a mother holding her firstborn. Or, in the world which is coming into being, is that point of view intolerably sentimental? Is it not better that the baby, and even the majority of its contemporaries, should die in infancy rather than grow up to the inevitable crowded misery and starvation of the whole population? Choices have to be made. By whom? On such issues the scientists have no more wisdom than the rest of us.

No wonder there are some voices calling for a moratorium on science. But that is an impossible solution. You might as well call for a moratorium on the growth of a twelve-year-old boy. The crisis of our time is not the advance of knowledge but the fact that we have *insufficient* knowledge and what knowledge we have has outrun our wisdom. As Tennyson wrote, "Knowledge comes, but wisdom lingers."

The accumulation of knowledge cannot be stopped, but the values to which research should be oriented have become, to an unprecedented extent, suddenly the responsibility of this generation. The scientist has been giving us knowledge with which he himself is quite unprepared to cope. Our social institutions, we ourselves, are not ready for this awful responsibility. That is why so many of us today long to turn the clock back. After all, who would not want to uninvent the atom bomb? Or undiscover what we and other nations can do in chemical-biological warfare? Are we completely helpless in the face of increasing knowledge? Are our political institutions able to cope with the dangers?

It may be that the Bill of Rights will have to be reinterpreted. Justice Holmes said that the right of free speech gave no man the right to cry "Fire!" in a crowded theatre. Shall we, for our social protection, say we have the equivalent of a crowded theatre in our dangerous world? Shall we say that, until we are assured that the knowledge acquired in a laboratory or elsewhere will be used for the enhancement of human life, we, through our government, shall exercise complete control, even veto power, over its dissemination? Or shall we allow engineers to change the face of our planet without our having complete knowledge of the side effects of that change? To give such powers of veto to the state looks pretty frightening, but what is the alternative?

I admit it frightens me. We have a deep-rooted and well-founded fear of a powerful government. Let me give you one example of what creates foreboding. In recent years we have learned how a government of a supposedly free people can use the means that science and technology have provided to inhibit personal liberties won at such great cost by our ancestors. I refer, of course, to the surveillance that various parts of the government, particularly the F.B.I. and the Army, are carrying on over individual citizens. We have assumed that every human being has a right to privacy, and the Supreme Court has thrown out many laws that would abrogate that right, such as those quaint laws passed in some states regulating the sexual relationship between husband and wife. In the past the attainment of privacy was comparatively easy. Since antiquity, third-party surveillance over private communications has been either through hidden physical presence, eavesdropping or surreptitious reading of correspondence. All easily avoided. In this century all that has changed. First came the invention of the telephone, and almost simultaneously came stories of telephone tapping to steal market quotations, to catch criminals, and to provide newspaper scoops. The telephone was followed by the dictagraph recorder by which, unknown to at least one of the conversers, recordings of conversations could be made. Then came the hidden mike. With this audience it is unnecessary to describe all the devices now used to collect information about people and the means of recording that information, some quite legal, some quite illegal, and some whose legality still puzzles the courts.

Information about persons is stored in a computer bank. These data banks have your credit rating, your contacts with the law, your income tax record, your fingerprints, your employment record, your medical history, and a great deal of just plain gossip about you. The fourth amendment says that "the right of the people to be secure in their persons, houses, papers, and effects against unreasonable searches and seizures shall not be violated." That section of the Bill of Rights is a recognition of civilized man's desire for a reasonable amount of privacy, and the desire of the individual to decide for himself how much publicity should surround his daily life. As one Philadelphia matron expressed it, "The public has a right to know that I was born, married, and died. Otherwise my life is private."

But our lives are no longer private. For an exhaustive and frightening account of how far we have gone in surveillance of each other, read Alan Westin's *Privacy and Freedom,* published in 1967, or Arthur Miller's *The Assault on Privacy,* just published. Since then even more scary bits of news have leaked out. The F.B.I. seems to have informers in every organization. The Army admits that it has dossiers on 25 million "personalities" and keeps data files on 760,000 organizations. The passport division of the State Department has a computer file on 243,135 persons who would not, according to Director Frances Knight, "reflect credit" on the United States. According to the *New York Times,* an airplane company has been asked by the F.B.I. where, and in whose company, its passengers travel. The F.B.I had scouts and informers to report on citizens who participated in Earth Day. That is nothing less than paranoia, and that paranoia has spread to other parts of the government, particularly the Army.

Assistant Attorney General Rehnquist says that the F.B.I. has a perfect right to put surveillance on a member of Congress, and further states that there is no need for legislation to limit the government's power to gather information about its citizens. All in the interests of national security. From the names of some of the men whom we know were spied on, we must gather that a statement of political difference constitutes in the eyes of these agencies a threat to our security.

Are we as individuals and citizens going to be timid in expressing ourselves for fear that a misinterpretation, or even a correct interpretation, of what we say will be fed into a computer to be examined and appraised years hence in a different cultural climate? If so, the Big Brother of George Orwell's *1984* is already here. It is good to have information about the criminal element in our society, but must its price be the destruction of individual privacy that for centuries has been regarded by western man as his precious right? There is certainly a justifiable distrust of big government, and when I am about to advocate big government I realize the dangers.

At present, we are adrift in history with terrible problems brought on by our own success. Figuratively, we are exploring new lands, but every promising haven seems filled with booby traps. As Berthold Brecht wrote, "Today every invention is received with a cry of triumph which soon turns to a cry of fear." The terrible fact of our society, as we face the future, is that the scientific and technological mind has raced far ahead of the political and social mind. New processes are being devised, new opportunities opened up, before institutions to deal with them can be developed. We are beginning to realize that science and technology make possible many things that we had better refrain from doing, that every advance has a potential for evil as great as, if not greater than, its potential for good. As was said of Hannibal, we know how to win victories but we don't know how to use them. To put it in another way: the great issue of our time is how to bring into better equilibrium the tremendous power that science has given men and the moral and political control of that power.

My own conviction is that this unprecedented and revolutionary situation in which mankind finds itself calls for unprecedented and revolutionary political measures if civilization is to get off a suicidal course. In the first place, we have to recognize that this is now one world. Marshall McLuhan has said that technology, particularly the development of instant communication, has made the world a global village. I think his figure of speech is inaccurate. In a village there is tradition and a tacit recognition of interdependence, of being one's brother's keeper. He had better have called it a global city, which, unlike the village, provides fragmented, temporary relationships and cannot give the sense of psychic wholeness that was so richly satisfying in village life. In a city, opposite interests have constantly to be reconciled, each interest having to yield some of its desires to make possible any community living. The growth of trade and international corporations has made national borders increasingly meaningless. The world is now a global city, and the existence of rival, national, sovereign states is an anachronism. So is war. Our leaders say that a war between the major powers is unthinkable, suicidal for both sides; but we and our alleged ideological rivals continue to prepare for it with a fantastic drain on the resources of both groups. I cannot see that the future of mankind is anything but precarious when in military matters we have a balance of terror, and when self-restraint of one nation in the use of scientific and technological power can be met by the unrestrained exploitation of that same power by another nation. We must act globally.

Can we have population control in the northern hemisphere and continued population explosion in the southern hemisphere? Can one nation pollute the ocean that is shared by all? We have become a global city and we must govern ourselves as a global city. We must work toward a strong world government, one which can recognize and even foster cultural differences, but one which has power to prevent international wars, and which can also prevent individual men and nations from destroying or abusing the limited resources of this spaceship, Earth, which is the only home our children will inhabit.

Nations must yield sovereignty just as individual businesses and institutions within a nation must abandon laissez-faire for more social control. It is romantic and nostalgic to look back on the individualism of the past, but old ways will not match the forces we have unleashed. Every society has found in time of danger that it was necessary to have a strong central government; and our society, with the potential for evil we have created, is going to live in danger from here on in. Individual men, institutions and nations must be prepared to yield power. One might almost say that for the longer-range future our choice is between one world and no world. Let me remind you that the course of civilization has been toward larger and larger political units, from tribe to nation. The United Nations must now become what its name implies.

Moral and political power must control the exploitation of scientific research. We are taking halting steps in that direction. Ecology has become a national and international issue. We are trying to control the use of insecticides that do far more harm than good, and we are attempting to deal with other health hazards. However, we are introducing new forces and new products at such a rapid rate and on such a wide scale that the effects come upon us before we have a chance to evaluate the long-range consequences. We know, after these many years, that the smog over our cities is caused by the automobile, which has become an indispensable part of our national life. The absorption of radioisotopes by the human body was known only after extensive testing. The resistance of our common detergents to bacteria became known only after they had become household necessities for many people. Drugs have been used before we have known their dangerous side effects. We can no longer afford that recklessness of ignorance. In introducing new factors into our national life, we have to test and then apply the concept of value to every scientific and technological advance, or we may learn too late that its exploitation can create more harm than good.

What I am saying is not that we must retard the advance of science and technology, but that we must stimulate our physical scientists to take the next step—which is to determine insofar as possible the ultimate social effect of the application of new knowledge. Scientists must work with social scientists on science-generated problems and clarify those problems so that political and ethical considerations can be applied intelligently. We need more science rather than less, but science with greatly broadened goals. Only after that further research has been carried out, and the cost of new knowledge has been measured, should the decision as to its use be made. Who is going to make those ultimate decisions? It will have to be a strong, bureaucratic, central government. They cannot be left to individual judgment, nor can they be determined by the normal democratic processes. Issues are now too complex and too important to be judged by men who are not well informed on the potentials for good and evil of any scientific breakthrough. We believe in democracy, or at least give lip service to that political ideal, but the complexity of modern life has not been hospitable to democracy. Take a simple issue facing us today, inflation. Can we put to a vote the fiscal policy this country should adopt to stop inflation? The question seems too complex for even our bright economists. And this is true of issue after issue; they are too complex for uninformed judgment.

So we have, and will have increasingly, the invasion of professionalism into political life. It is the professionals who work on economic planning, public health, and the balance of payments. The voters may elect line officers, but the officers are increasingly dependent on their staffs. Through democracy we can determine the broad overall goals of our society, but not the techniques by which they are reached. Just as, in our economic life, it is the managers who run our corporations and not the stockholders, so in our political life we are going to be increasingly governed by the professionals or, if you want to use the term, the bureaucrats. We are all aware of the stultifying effect that bureaucracy can have, but I see no other solution. Centralized planning is necessary to our survival. I know that only a few minutes ago I illustrated the dangers of a strong government and how it can abuse its powers. It frightens me that a government should have so much power, but is there an alternative? The counterforce will have to be an educated and alert citizenry, directing its constant criticism at our elected leaders. It may well be, as H. G. Wells wrote over fifty years ago, that the history of the world has become the story of a race between education and catastrophe. Continuous, lifelong education.

So we have to have a stronger national government, and we have to abandon tribal nationalism in favor of world government. These governments, through their professional staffs, must apply the yardstick of short-term and long-term value to every new possibility that science offers man. These values themselves have constantly to be reappraised. We know now that man cannot for much longer be a parasite on earth, exploiting nature for his immediate ends regardless of consequences. He will have to work in harmony with the processes of nature, preserving our wild life and wilderness, and to the maximum extent possible, recycling our limited raw materials for further use.

Above all, I believe that we as a people have to redefine what we mean when we use the phrase Gross National Product. Hitherto we have meant the creation of goods and services. That yardstick must change. We must shift from emphasis on the "goods" life to emphasis on the good life. Instead of the production of steel or even the increase in telephone circuits, we should emphasize the enlargement of educational opportunities, the increase in standards of health and nutrition, the increase in wild life, the growing cleanliness of our lakes and rivers, the lifting of cultural standards of the press and TV. If achieving such a GNP means little or no economic growth in the old terms, the price is petty. As Erich Fromm has written: "Man, not technique, must become the ultimate source of value; optional human development, not maximal production, the criterion for all planning."

In closing let me say that for good or ill, mankind has at last achieved dominion over the earth, and with that power has suddenly also inherited an awesome responsibility. For the first time in history we have the knowledge and power to destroy ourselves and the beautiful spaceship we inhabit. For the first time in history we have the knowledge and power to free mankind from cold, hunger, plague and ignorance. History indicates that men have tended to use power more for evil than for good. But in the past, when men had limited power, mistakes were retrievable. A major mistake now can be irretrievable. In this generation and the next, the human race has to make a leap toward maturity that will enable us to create a new social order based on decent values. It is that or Armageddon, in whatever form it may take. In his *Study of History*, Arnold Toynbee pointed out that in each of the previous twenty-one civilizations there came a challenge to which the civilization was unable to respond, and that civilization disappeared. None, however, faced so awesome a challenge as do we, the twenty-second. If our civilization can respond, the future is unlimited. If not, there may never be a twenty-third.

ARTHUR E. JENSEN

Arthur E. Jensen, a member of the Dartmouth College faculty for more than thirty years, is a graduate of Phillips (Andover) Academy and Brown University. Upon graduation from Brown in 1926 he joined its English department as instructor. He left Brown to study at the University of Edinburgh in Scotland, where he took his Ph.D. in 1933. That year he joined the English department at the University of Maine. While in Maine he served as vice chairman of the Maine Institute of World Affairs. He joined the faculty at Dartmouth in 1937.

During World War II, he served as coordinator of the Civilian Pilot Training Program centered at Dartmouth. When that program was absorbed by the Navy, he accepted a commission in Naval Aviation. When the V-12 program was introduced, he served as commanding officer of four separate V-12 Naval Stations. After the war he returned to Dartmouth as a professor of English.

He became director and chairman of the Great Issues Course and in 1951 became chairman of the department of English. In 1954 he became dean of the faculty, retiring from that post in 1964 to rejoin the department of English and to continue his outside activities.

In the summers of 1956 through 1961 he was director of an eight-week Conference in Liberal Arts for Executives held at Dartmouth. From 1957 through 1961 he was also director of a similar conference for executives of the National Association of Mutual Savings Banks, and since 1959 he has been director of a Conference on Management Objectives for executives of the American Telephone and Telegraph Company. He has been a consultant and program leader for the Management School of I.B.M., for Nationwide Insurance Company and for Western Electric.

He is the author of several articles in professional and other journals. He has three honorary degrees: L.H.D. from Brown, Lit.D. from Long Island University, and LL.D. from Windham College.

10. REVOLUTIONARY PHILOSOPHY OF THE SEVENTIES

FULTON J. SHEEN
Archbishop
Roman Catholic Church

I thought that instead of talking about business and ethics and such things, I would try to give you good people a view of a basic philosophy and thought that is governing our world, and is going to govern it for the next fifty years, at least.

I'm going to talk about violence. We live in a revolutionary age. Maybe it began in 1917, when Marie Alexandrovitch was teaching a group of children in one of the churches in Moscow. Horsemen came in and destroyed the statuary and the altars, then charged the children. Marie Alexandrovitch ran out and went to Lenin, whom she knew, and told him what had happened. He said, "Yes, I sent them."

The revolution was on.

Now, what makes this a revolutionary age? Here I'm not just speaking of a political revolution or an economic revolution. It's the spirit that pervades people.

There are several reasons for violence in any age. One is the loss of meaning and purpose. Just as soon as mankind loses a sense of the goal of life, man begins to be revolutionary. If a boiler, for example, were conscious and said to itself, "Why should I retain steam at a given pressure and obey a law imposed on me by an engineer? I'm going to revolt"—the boiler would explode. It would have lost its meaning.

If a train said to itself, "Why should I follow these tracks that were laid out by someone before my time," and it jumped the tracks, it would no longer be free to be a train.

If we became revolutionary and decided to draw a triangle with thirty sides instead of three, we would lose the freedom to draw triangles.

And so as the goal and purpose of life is lost, men become violent. And there has been creeping into drama, in particular, this sense of revolt. In France, Sartre said: "I cannot have anyone outside ever give me a goal or a purpose or a destiny, because if he's outside of me he sets a limitation on me." And so in his play called *No Exit*, which I believe showed here in New York, there were four characters in Hell, and each one talked about his ills, his complaints, his hates. No one listened to the other. They were only anxious to deliver themselves of their own fears and egotisms.

The last line of the play was: "'My neighbor is Hell, because my neighbor sets a limitation to me. He therefore restricts my freedom. And I must revolt against him." Then the curtain fell.

Now, you can work this out in your own mind, with all that you know about revolution. I'm merely trying to give you a basic reason why it exists.

A second reason why it exists is the general influence of Marxism in the world. We talk a great deal about communism, but very few know the philosophy of it. I taught it for about ten years. I read every book that Stalin ever wrote. He published about sixteen volumes (few would believe that he wrote a work on "criteriology"); and I've read Lenin and Marx.

And so I'm going to try to show you how this philosophy of Marx will have a bearing on your business and your company. Marx says that man has been alienated from himself, that he no longer follows his true nature. The destructive forces have been religion on the one hand, and private property on the other. Religion, first of all, destroys a man's true nature because it makes him subservient to God. Private property destroys man's nature because it makes him subservient to an employer. If, therefore, man is ever to be completely free, there must be the destruction of religion and the destruction of private property. As you see, the two principles go together. If you have one without the other, man is still half enslaved.

We've gotten away from the crude dialectics of Marx in the course of time, but the revolutionary attitude still prevails. Putting the two reasons for revolution together, you have, first of all, an affirmation of the human ego and the denial of any limits, objectives, norms or standards outside of himself. That makes a man impatient of limitation. If, for example, I am the supreme master, I decide that a yard is going to be forty-two inches long. You say it's thirty-six, but you're using an objective norm, not a subjective norm. I am the determinator of what is right.

You can even see how some modern art is related to this approach.

Sometimes you can't tell whether you're looking at a picture of a drunken man going up a stairs, or a stairs going up a drunken man. Why can't you? Because you're looking for some norm, an object. You say, "Well, that's supposed to be orange. Where's the orange?" Here's the denial of the object.

So it's the subject that determines everything. That's the egotistic side of revolt.

The other side of it is the Marxist side, which is being taught in this country by Henry Marcuse. Henry Marcuse is the philosopher of many young students. According to him, society is so corrupt that, though you may never be able to attain your objectives, you must constantly keep the Establishment off balance by a series of successive and protracted revolutions.

Now, this is the philosophy that's governing much of the world. Some of it is in the open; some of it is hidden.

Is this revolutionary spirit American? One of the young revolutionists of the United States said that "Revolution is just as American as cherry pie." The President's Commission on Violence held that revolution is not cherry pie to Americans, but that dissent and protest are—which, indeed, is much more true.

But the young revolutionists are saying: "We began with revolution. Therefore, we are a revolutionary people." Is that true? No, it is not true.

The French Revolution had some relation to ours. Thomas Jefferson was in France at that time. In the beginning, the French Revolution was presided over by Saint-Just, who held that you must completely overthrow any order which exists. Jefferson said to Saint-Just: "Do not do this. Nature does not suggest it. Nature suggests a slow evolution of growth, not the complete destruction of foundations." Jefferson also differed from Saint-Just in holding that in a political society you have relativity; not everyone accepts the same view of things, and they should not; we have to agree to tolerate different parties and different points of view.

What was the French idea? The French idea was: You must transport mysticism. Mysticism belongs to theology. Mysticism is that form of spiritual life whereby you live to be completely identified with God. In other words, it's a very intense form of sanctity. And Jefferson said to Saint-Just: "What you're doing is taking from Heaven, from theology, the absolute commitment to God, and you're bringing it down to the political order, which is wrong."

So in answer to our young revolutionists, we are not revolutionary in the sense that they contend. On the contrary, when we wrote our Declaration of Independence we had to search about for the source of our rights and liberties. Where, for example, do I get the right of free speech? Where do you get the right of assembly? From the state of New York? If we got these from the state of New York, the state of New York could take them away. Did we get them from the federal government? If we got them from the federal government, the federal government could take them away. It's taking almost everything else away, it could take these away too.

So our founding fathers asked themselves the question: "Where do we get rights and liberties?" From the will of the majority? No, if rights and liberties come from the will of the majority, the will of the majority could take away the will of the minority.

And so they set it down in the second paragraph of the Declaration of Independence that it is a self-evident principle that the Creator —the Creator—has endowed man with certain inalienable rights. In other words, we get our rights from God, not from the State, not from Congress, not from the majority. Therefore no one can take them away. That is why Amendment Nine of our Bill of Rights states that when any rights are mentioned by the Constitution, it must never be assumed that the people have no other rights than those granted by the Constitution.

Up to this point, then, we have in the world a developing philosophy of revolution. And, incidentally, every carnal age in history, every age with a strong emphasis on sex, has always been a revolutionary age. You can verify that in Sorokin's four volumes on *Social and Cultural Dynamics,* so I will not go into it. But it is a kind of sociological cause.

So on the one hand we have this revolutionary philosophy in the world. We have also the contention that it is American. I'm contending it is not American. What they're talking about is the French Revolution of Saint-Just, not the American Revolution of Thomas Jefferson.

How is it going to affect us? How will it affect you? You will have protestors within ten years, not from inside your organization but from outside.

Students' dissent will move from universities, has moved from universities, moved against governments. It will move against corporations. Their argument against corporations does not have much validity, but they are looking for new forms of protest. They speak of the impersonal character of ownership. It is an attack for which you must be prepared. I'm talking about the future, about something that is going to come. You're almost the only area that it hasn't touched. I've had it. You'd think that I would be immune.

For example, I was shocked by conditions that I found in Rochester, and by the general apathy about doing anything for the poor. There were indescribable conditions of poverty—indescribable. Boys and girls who have never slept in a bed in their lives, who never had a knife and fork in their hands and who stay up much of the night battling rats.

Then there is the attack against the Church on account of its wealth. I wish I had known where some of it was when I was running a diocese. I would have loved to draw on it. We've got a lot of old buildings which we can't keep up.

What I attempted to do was to give away a church. It was not just for the sake of Rochester. I wanted to establish a beachhead that would develop housing for the poor. And I felt the way to do it was to give the government some land with which to start, and then they could reach out through the inner cities. I began with Rochester because I had the power to do it there, but I wanted it done also in Chicago, New York, Philadelphia, Pittsburgh, Los Angeles, San Francisco—everywhere. So I went to Mr. Weaver who was the head of housing in Mr. Johnson's cabinet, and I proposed giving him any property that he wanted. I said, "In order that you may realize that we're not giving something we can't use, that we will not use, or that we have no use for, you send representatives to Rochester and take any church that you want—church, house, school, anything— just so you use it for building."

So they sent four men up from Washington, and they selected a parish. Mr. Weaver asked that it be kept secret. I contacted the proper authorities to alienate church property, both in the diocese and outside, and when finally Mr. Weaver and I made the announcement, I had protests, placards. Someone went out to a girls' college, loaded six girls into a car, brought them down to our office and put placards in their hands. Then they called the TV stations, who are always ready to picture a protest, or someone being hit, so that I finally had to withdraw the proposal.

I'm only using this as an instance to show that today nothing is sacred—nothing. How soon it will be before it touches you, I do not know, but you must anticipate it.

How can it be anticipated? You are a tremendous organization, national and beyond, I will tell you something about their revolutionary thinking, what their arguments are, and from this you can develop plans.

Their argument is, principally, that great corporations are far more interested in profits, or returns on investments—stock for their stockholders—than they are in the good of the community. That's their argument. Their protest will take that form.

It's very difficult for a great national organization to decide how it can help the community, because you're in every community, almost in every home. But I do think that one of the ways may be to take a greater interest in communities, and not just in profit.

I know that when I went to Rochester, I told the chamber of commerce that the social conditions existing in the city were really, to the great brains that were there, nothing but a pimple on a nose.

The corporations of this country could clean up the social problem in the United States in four years. You have the know-how, you do not have to go through government red tape, you can solve almost any problem that has to do with Matter. And as I look into the future, I see an escape from increasing government control by corporations and industries becoming far more concerned about the communities than they have been.

Xerox asked me to do an introduction to a film. We went to a street which was in a very poor section of the city—though not the poorest by any odds—and they set up a television camera. As soon as I started talking, about forty black kids rushed toward me. They knew me. It was all very friendly, and I was introducing my subject in an off-hand way. I said: "When we have visitors to this city, why is it that we say to the visitors, 'I'm sorry to bring you down this street'?" I said: "Why do we say that?" It was only a rhetorical question, but one thing you never do when you're talking to kids is ask a rhetorical question. They always answer.

So one little kid said: "Because they don't want to see our rats" and I said: "Come here." I put my arm around him and finished the film sequence, which never appeared on TV. Afterwards I said to him: "What is your name?" He said: "Calvin Gordon." I said: "I'm going home with you." So I went home with him and I was taken into two rooms. It was an old house, maybe originally a seven- or eight-room frame house, and it was now broken up into seven apartments. This family of three boys, mother and father, lived in these two rooms; no water, dark, dingy, foul.

And so I said to Mrs. Gordon, "How much do you pay here?"

She said, "Seventy-four dollars a month—to a lawyer." But, she said, "What can we do? We can barely exist, we will never have a down payment on a house." So I said, "Let's go out and find a house." So we walked around the better neighborhoods and found a for-sale sign. I called the real estate man. I said, "How much do you want for the house?"

He said, "Ten thousand dollars."

I said, "I'll give you eight."

"Oh," he said, "wouldn't consider it."

"Well," I said, "come and see me."

When I saw him I said, "I'm buying this house for a Negro family. Would you sell it to me for eight?"

He said, "Yes."

"Will you give me some furniture?"

"Yes, furniture for two rooms."

So I bought the house. I only made the down payment on it.

Then we got an FHA loan, and in fifteen years they will own the house. With insurance and taxes, the monthly payments are $60.

That started a housing fund. So that by the time I left Rochester we had bought nineteen houses.

Industry could become interested in this sort of thing, and on a bigger scale. I am merely suggesting that you be prepared not for a coming revolution, nothing that big, but for this revolutionary spirit that I have been describing. It's in the air. It's against everything. The common word today is "Down"; a word you never hear is "Up." We live in a world of protest, not of affirmations, a world of doubts. As Gazey said, "Don't be telling me your doubts. I've got doubts of my own. Give me some certitudes."

I was invited this morning to appear on a television show in which there were to be eight or ten other men. I don't remember what the subject of discussion was, but I said "No." All kinds of opinions were to be represented on the panel. I said "Why don't you pick out someone, just one of them, and give the people a few certitudes in these days of doubts?" Where will our certitudes come from? Where will the change come from? Are we in a hopeless situation? We certainly are not, not by any stretch of the imagination.

You know, the hope of the world may come from the eastern parts of Europe. A change is going on there of which we seldom hear. Did you know, for example, that the Russian soldiers who came into Czechoslovakia hid their families in their tanks? They were looking for a better homeland.

Some of the most hopeful thought in the world is coming from eastern Europe. They've already tried the revolutionary spirit.

Jurgon Moltman, who is a Lutheran theologian, said that he went to Prague to attend a Communist congress, and while at the Prague airport he bought a copy of *Time*. The lead article on the cover was: "God is Dead." When he got to the Communist congress, what was the title of their discussion? God is not quite dead. Some months later he went to a congress at Marienbad, Czechoslovakia. There were Protestants, Catholics and Communists holding common discussion. All the Protestants and all the Catholics spoke on one theme: the necessity of getting religion involved in the world. What did the Communists talk about? The transcendence of God. They said: Our political and our economic systems are closing us in. We need some hope. We've got to break this egg, this shell. That was their theme. You would have thought it would be the other way around.

Will there be fulfilled what Dostoyevsky, the great Russian novelist, wrote in the last century, when he said that Russia would one day become filled with devils? And then he said, to someone reading the Gospel of Luke about the young man in the land of the Gadarenes: That's my Russia, my beloved land, full of devils. One day, he said, the devils will be driven out of Russia. They'll be driven into the swine. The swine will be pushed back and back and back into the of Christ and learn his Gospel.

Here was a man who was in Siberia, who wrote about communism sea and there they will be drowned. And Russia will sit at the feet as if it existed in his own time, and this is what he foresaw.

Hope can come from unexpected places, and hope can come from here. With this note I will conclude.

I've been talking about violence, what causes it, whether or not it is American, how eastern Europe is turning away from the philosophy of revolution, and will, I think, turn more and more away from it.

And what about ourselves? We have to introduce violence into our national life too. There are two kinds of violence. One is the sword that's outward, the sword that Peter used in the Garden, the sword that swings and hacks off the ear of the servant of a high priest, that cuts into a neighbor.

The other sword is the sword that our Blessed Lord spoke about: "I came not to bring peace, but a sword." It's the sword that cuts inwards at ourselves, our vices, our concupiscences, our selfishness, our materialism.

All kinds of mortification, self-discipline, self-restraint are passing out of our national life. And as this internal violence ceases, external violence increases. About the only educational institutions where self-restraint is practiced are Annapolis, West Point, the Air Force Academies. We're even losing it in our seminaries.

And somehow or other we must begin to meet violence with violence. We must think of a great lady who sang the most revolutionary song that was ever written, the *Magnificat*. My, she was violent: "The rich, you will send away empty. You will disperse the proud. You will rid those who are already in possession."

She spoke a revolutionary hymn. It was a revolutionary hymn that began first with self, and that in a certain sense was the meaning of Christmas. As soon as He came, He brought violence. There was violence against Him. There was the sword immediately, so that He had to flee into Egypt.

But that external, outward violence He never used against anyone. And sometime, somewhere, maybe we who are supposed to be the bearers of that Gospel will not be quite so anxious to identify ourselves with the revolutionists who are concerned only about cutting off their neighbors' ears. Maybe we'll begin to preach the right kind of violence to which we were called, the violence to ourselves that will calm this outward violence and will make America what it was destined originally to be: a land of peace and prosperity. Thank you.

FULTON J. SHEEN

Archbishop Sheen is probably the best-known Catholic prelate in the nation. His radio program "The Catholic Hour," which began in 1930, ran for twenty years and by 1950 was carried worldwide. His television program "Life is Worth Living" appeared weekly from 1951 to 1957. Carried on 123 stations, and estimated to reach

30,000,000 people each week, the program received an Emmy award in 1952 and the *Look* magazine award for three successive years. In 1952 Bishop Sheen also received an award from the Forensic Society as an "outstanding speaker in the field of religion."

Archbishop Sheen received his undergraduate and master's degrees from St. Viators College in Bourbonnais, Illinois, and completed his theological studies there and at St. Paul's Seminary in St. Paul, Minnesota.

He earned advanced degrees in theology at the Catholic University in America and a doctorate in philosophy at the University of Louvain in Belgium. He was awarded the much coveted "Agrégé en Philosophie" by Louvain University in 1925. He also studied at the Sorbonne in Paris and the Collegium Angelicum in Rome, and has been presented with innumerable honorary doctoral degrees and educational citations.

His religious and pastoral assignments within the Catholic church have been extensive. For twenty-three years he taught philosophy at the Catholic University in America, a professorship he resigned in 1950. In 1934 he was appointed papal chamberlain and a year later was elevated to domestic prelate. In 1951 he was nominated bishop by Pope Pius II, and consecrated in Rome. The archbishop has served in various church offices; among the most important was the national directorship of the Society for the Propagation of the Faith, the church's principal mission organization, which he held from 1950 to 1966. In 1966 he was appointed bishop of Rochester, New York. He resigned that position three years later and was appointed archbishop of the Titular See of Newport, Wales, a position he holds today.

Despite a busy schedule, the archbishop has found time to become a prolific writer. He produces the weekly newspaper column, "Bishop Sheen Writes," for the secular press, and was editor of two religious publications, *World Mission* and *Mission* magazines. During his career he has written 66 books, as well as numerous articles which have appeared in many publications.

In his writings, as in his popular radio and television programs, his themes cover the subjects of ethics, morality, the evils of communism, the improvement of the quality of life within America, and contemporary theological and philosophical developments in Europe and America.

Archbishop Sheen has given extensive spiritual conferences to the clergy in the United States and Europe. He has lectured at most of the leading universities and colleges of the United States, including West Point and Annapolis.

BIBLIOGRAPHY

Because the complete list of books written by Archbishop Sheen is extensive, this bibliography includes only books listed in 1971 as still in print.

Moral Universe: A Preface to Christian Living. Freeport, N.Y.: Books for Libraries, Inc., 1936.

Preface to Religion. New York: P. J. Kenedy & Sons, 1946.

Three to Get Married. New York: Hawthorn Books, Inc., 1951.

Peace of Soul. Garden City, N.Y.: Doubleday & Co., 1954.

Lift up Your Heart. Garden City, N.Y.: Doubleday & Co., 1955.

Life of Christ. New York: McGraw-Hill Book Co., 1958.

Go to Heaven. New York: McGraw-Hill Book Co., 1960.

A Priest Is Not His Own. New York: McGraw-Hill Book Co., 1963.

Power of Love. New York: Simon & Schuster, 1965.

Footprints in a Darkened Forest. New York: Hawthorn Books, Inc., 1967.

Guide to Contentment. New York: Simon & Schuster, 1967.

ed. *That Tremendous Love.* New York: Harper & Row, Publishers, Inc., 1967.

Children and Parents. New York: Simon & Schuster, 1970.

Moods and Truths. Port Washington, N.Y.: Kennikat Press, Inc., 1970.

Old Errors and New Labels. Port Washington, N.Y.: Kennikat Press, Inc., 1970.

Science, Psychiatry and Religion. New York: Dell Publishing Co., n.d.

These Are the Sacraments. New York: Doubleday & Co., n.d.

Way to Happiness. New York: Crest Publishing Co., n.d.